T0235767

The Presence of Nature

The Presence of Nature

A Study in Phenomenology and Environmental Philosophy

Simon P. James
Durham University, UK

First published 2009 by
PALGRAVE MACMILLAN

Palgrave Macmillan in the UK is an imprint of Macmillan Publishers Limited,
registered in England, company number 785998, of Houndmills, Basingstoke,
Hampshire RG21 6XS.

Palgrave Macmillan in the US is a division of St Martin's Press LLC,
175 Fifth Avenue, New York, NY 10010.

Palgrave Macmillan is the global academic imprint of the above companies
and has companies and representatives throughout the world.

Palgrave® and Macmillan® are registered trademarks in the United States,
the United Kingdom, Europe and other countries

ISBN 978-1-349-30792-0   ISBN 978-0-230-24852-6 (eBook)
DOI 10.1057/9780230248526

This book is printed on paper suitable for recycling and made from fully
managed and sustained forest sources. Logging, pulping and manufacturing
processes are expected to conform to the environmental regulations of the
country of origin.

A catalogue record for this book is available from the British Library.

A catalog record for this book is available from the Library of Congress.

10 9 8 7 6 5 4 3 2 1
18 17 16 15 14 13 12 11 10 09

Contents

Acknowledgements

In writing this book, I have benefited from the help of many friends and colleagues. I am especially grateful to Alex Carruth for reading through a draft of the entire manuscript and for making a number of suggestions that greatly improved the finished product. But I would also like to extend special thanks to David E. Cooper for his encouragement and advice, and for the many, characteristically perceptive comments he made on several draft chapters. I am indebted to Matthew Ratcliffe, Helen Curry, Alison Stone, Wayne Martin, Charles Brown, Wolfram Hinzen, Liz McKinnell, Amanda Taylor and Ian Kidd for the feedback they provided on pieces of work that eventually made their way into the final manuscript, and I am also grateful to Priyanka Gibbons, at Palgrave Macmillan, for the help and advice she has given me throughout the writing process.

Material from this book has been aired at talks I have given at the universities of Durham, Lancaster and Liverpool, at conferences in Oregon, Colorado and Dundee, and at workshops at the universities of Durham and Edinburgh. I would like to thank the audiences at these events for providing me with so much food for thought.

Some of the material in this book has been published elsewhere. I am grateful to White Horse Press, the publishers of *Environmental Values*, for granting permission to reprint parts of my article 'Phenomenology and the Problem of Animal Minds' in Chapter 2. Moreover, Chapter 5 is loosely based on my article 'Merleau-Ponty, Metaphysical Realism and the Natural World', and I would like to thank Maria Baghramian, the editor of the *International Journal of Philosophical Studies,* for granting permission to reprint this material.

Finally, I would like to extend a heartfelt thanks to Mum and Carole for all those Saturday and Sunday morning walks with Winston and Lucy on Puttenham Common. If I hadn't had a chance to spend so much time out in nature as a child, I doubt I would have had so much time for it as an adult.

Abbreviations

Works by Husserl

CES   *The Crisis of the European Sciences and Transcendental Phenomenology: An Introduction to Phenomenological Philosophy*, trans. D. Carr (Evanston: Northwestern University Press, 1970).

Works by Heidegger

BT   *Being and Time*, trans. J. Macquarrie and E. Robinson (Oxford: Blackwell, 1997).

BW   *Basic Writings*, D. F. Krell (ed.) (Oxford: Blackwell, 1996).

CP   *Contributions to Philosophy (From Enowning)*, trans. P. Emad and K. Maly (Bloomington: Indiana University Press, 1999).

DT   *Discourse on Thinking*, trans. J. M. Anderson and E. H. Freund (NY: Harper & Row, 1966).

FCM   *The Fundamental Concepts of Metaphysics: World, Finitude, Solitude*, trans. W. McNeill and N. Walker (Bloomington: Indiana University Press, 1995).

IM   *An Introduction to Metaphysics*, trans. R. Mannheim (New Haven: Yale University Press, 1959).

N   *Nietzsche*, trans. F. A. Capuzzi (San Francisco: Harper & Row, 1979–87). (Four volumes: N1, N2, etc.)

Pa   *Pathmarks*, trans. W. McNeill (Cambridge: Cambridge University Press, 1998).

O   *Ontology – The Hermeneutics of Facticity*, trans. J. van Buren (Bloomington: Indiana University Press, 1999).

PLT   *Poetry, Language, Thought*, trans. A. Hofstadter (NY: Harper & Row, 1971).

QCT   *The Question Concerning Technology and Other Essays*, trans. W. Lovitt (NY: Harper & Row, 1977).

Works by Sartre

BN   *Being and Nothingness*, trans. H. E. Barnes
      (London: Routledge, 1991).

Works by Merleau-Ponty

BWr   *Basic Writings*, T. Baldwin (ed.) (London: Routledge, 2004).

EP   *In Praise of Philosophy*, trans. J. Wild and J. Edie (Evanston:
      Northwestern University Press, 1970).

N   *Nature: Course Notes from the Collège de France*,
      compiled by D. Seglard, trans. R. Vallier
      (Evanston: Northwestern University Press, 2003).

PP   *Phenomenology of Perception*, trans. C. Smith
      (London: Routledge, 1996).

PrP   *The Primacy of Perception and Other Essays on
      Phenomenological Psychology, the Philosophy of Art, History
      and Politics*, trans. J. M. Edie (Evanston: Northwestern
      University Press, 1964).

S   *Signs*, trans. R. C. McCleary (Evanston: Northwestern
      University Press, 1964).

SB   *The Structure of Behaviour*, trans. A. L. Fisher
      (London: Methuen, 1963).

SNS   *Sense and Non-Sense*, trans. H. L. Dreyfus and
      P. A. Dreyfus (Evanston: Northwestern University Press,
      1964).

VI   *The Visible and the Invisible*, trans. A. Lingis (Evanston:
      Northwestern University Press, 1968).

WP   *The World of Perception*, trans. O. Davis (Routledge:
      London, 2004).

Introduction

Environmental philosophers have had a great deal to say about nature as it has been conceived by scientists. They have had much to say about nature as it has been conceived by other philosophers. But they have had comparatively little to say about nature as we experience it in the living of our lives. In the following, I argue that this lack of attention to nature-as-experienced is a cause for regret, and not just for rhetorical reasons, not just because it makes the discipline of environmental philosophy seem disagreeably abstract and high-flown. I contend that this inattention to experience is a bad thing because environmental philosophy that fails to connect with our lived experience of nature is, more often than not, bad philosophy.

If the first general aim of this book is to challenge a popular, overly abstract approach to environmental philosophy, its second is to show, through example, the merits of an alternative approach. My second general aim, then, is to do environmental philosophy by paying close attention – closer than is usual – to how we experience the natural world. Accordingly, I engage not just with the works of philosophers, but also with the testimonies of a diverse collection of naturalists, scientists, poets and explorers – from the writings of J. A. Baker to the poetry of William Wordsworth, and from the passionate prose of Henry David Thoreau to the more sober reflections of modern-day scientists.

Taking such an experience-focused approach, I develop original accounts of (1) what the natural world *is*, and (2) how we *ought* to act towards it. And along the way, I hope to cast new light on some of the key problems in environmental philosophy: What is our place in nature? Do any non-human animals have minds? How are we to conceive our moral relations with the natural world? Does the natural world exist independently of our understanding of it? These are all familiar topics, of course, but this will be environmental philosophy seen from the ground up, starting from our lived experience of nature.

Φ

If these are the book's aims, its subject matter is the natural world as we experience it in the living of our lives. So this book is about the many faces of nature-as-experienced. It is about nature's familiarity and what Iris Murdoch once called its 'sheer alien pointless independent existence' (2007: 83). It is about those parts of the natural world to which we find ourselves drawn, and those from which we feel repelled. In the most general terms, it is an attempt to understand what it is like to inhabit a world that is in various senses and to varying degrees 'natural' rather than 'human' – a world of roots, soil and leaves, and not just of plastic and tarmac.

Expressed in more technical language, this book is about 'the phenomenology' of our relations with the natural world. Yet the word 'phenomenology' denotes not just the object of this study but also its method, for in examining what it is like to experience the natural world I will adopt a phenomenological approach.

I will have more to say in a moment about what it means to adopt a phenomenological approach to our relations with the natural world (and in particular, I will clarify what I mean by 'nature' and 'the natural world'). First, however, I ought to say a few words about what it means to adopt a phenomenological approach to anything.

In very general terms, to take a phenomenological approach is to do philosophy by attending to and reflecting on one's experience. At first sight, this might seem an odd way to go about doing philosophy. After all, it is commonly thought that it is not the philosopher's job to reflect on what it is like to experience phenomena. To be sure, some cursory observations about how things present themselves to us in experience may provide a fitting hors d'oeuvre, but it is often supposed that, once she has made these observations, the philosopher should proceed fairly quickly to the main course, a set of reflections on more abstract and general topics. So, for example, having briefly registered the fact that she seems to be surrounded by various material objects, the philosopher should feel perfectly justified in moving on to consider the more interesting question of whether those objects are really there.

Or having made some quick observations about the apparently conscious behaviour of her fellow humans, she should feel free to consider the meatier philosophical question of whether those apparently conscious beings really are conscious and not mindless automata or 'zombies'. This is simply the vaguely Platonic way in which philosophical inquiry is thought to proceed: away from the confusion and ambiguity of the phenomenal world and, as quickly as possible, onwards and upwards towards a clearer, cleaner realm of pure abstraction.

The worry, then, is that to the extent that the phenomenologist focuses on experience, she won't be doing anything philosophically interesting, and that to the extent that she does anything philosophically interesting, she won't be focusing on experience. But this worry is unfounded (as is the general conception of philosophy on which it rests). For it is a mistake to suppose that describing what presents itself to us in experience is an easy task and one that can (and should) be quickly got out of the way before proceeding to the main business of doing philosophy. On the contrary, as one phenomenologist points out, it requires 'much time and effort' to lay bare 'the world which is revealed to us by our senses and in everyday life' (*WP*: 39). Time and effort are needed because we – philosophers and to some extent non-philosophers too – are subject to certain entrenched prejudices which lead us to misconstrue what presents itself in experience. This is not to say that we tend to misunderstand *why* things present themselves to us in experience in the way they do (although we are no doubt prone to do this). It is to say that we tend to misconstrue *what* presents itself to us in experience. That which seems so clear and obvious, the world revealed to us by our senses and in everyday life, is not what we think it is at all. So, for instance, it is often supposed that experience presents us with what Maurice Merleau-Ponty calls an 'objective world', a realm of determinate objects arrayed in Euclidean space according to an absolute and quantifiable measure of time, and related to one another by various external relations. But does the world really present itself to us in this way? It might seem obvious that it does, but to be certain, one way or the other, one must interrogate one's experience. And to interrogate one's experience, or at least to do it thoroughly, one needs to employ a phenomenological approach. In order to discover 'the phenomenology', one needs to do phenomenology. To do phenomenology is not merely to catalogue the content of experience. For one thing, phenomenologists are not primarily concerned with *what* one experiences, but with *how* one experiences it. Their primary concern is not with the object of experience but with one's experience of it. And here their chief aim is, by attending to and reflecting on what presents itself to us in experience, to discover general truths about how anything is experienced. This is philosophically significant in two respects. First, phenomenological inquiries can have a deconstructive or therapeutic effect. In particular, they can (and have been used to) undercut a variety of pernicious sceptical doubts. So, for instance, a phenomenological approach can be used to undercut sceptical doubts about the reality of the external world or the existence of other minds. Second, and more positively, phenomenological inquiries can yield new and interesting philosophical insights. In *Being and Time*, Martin Heidegger uses a phenomenological approach to investigate what it means for anything to be (his famous 'question of the meaning of Being'). Through careful reflection on what presents itself to us in experience, Merleau-Ponty develops an original conception of what it means to be embodied. Proficiently employed, the tools of phenomenology can be used to shed light on the deepest and most important issues in philosophy. So if phenomenological inquiries can dispel a variety of philosophical confusions, they can also open our eyes to new philosophical truths. Either way, when it is done well, phenomenology can transform the conceptual landscape of philosophy.

Φ

I have been referring to how we experience the world, but of course all this applies to the *natural* world as well. In fact, interest in environmental phenomenology or 'eco-phenomenology' has been rapidly growing, with the result that it is now a recognised genre in the environmental literature. This rise in interest owes in large part to the influence of a few treatises. Here several works could be mentioned: Erazim Kohák's *The Embers and the Stars*, a beautiful paean for our lived experience of natural things, inspired chiefly by the works of Edmund Husserl; Lorne Neil Evernden's *The Natural Alien*, a critique of technocratic, managerial approaches to environmental problems and a sustained argument for an approach that takes seriously our pre-reflective encounters with nature; David Abram's *The Spell of the Sensuous*, a call, inspired by the works of Merleau-Ponty, for us to reacquaint ourselves with what it is like to inhabit the natural world 'from within' (1996: 65). The list could be extended, but there is no need: we will engage with many of the seminal treatises in environmental phenomenology in the following chapters. For the moment, I would like instead to consider what exactly it means to do *environmental* phenomenology.

The first thing to note is that environmental phenomenologists seek to elucidate what is often referred to as our *lived* experience of nature, nature as it discloses itself to us in the living of our lives (see Kohák 1984: 22). To do this, they set themselves the task of bracketing or putting out of play certain entrenched second-order or theoretical conceptions of what nature is, notably the widespread assumption that the world is not only an objective world, in Merleau-Ponty's sense, but also one that is fundamentally material in composition. As one environmental phenomenologist explains:

> Far more than we ourselves usually realize, when we make seemingly obvious assertions about 'nature', we are no longer speaking about the natural environment of our lived experience. ... Our statements are far more likely to refer to a highly sophisticated construct, say, matter in motion, ordered by efficient causality, which is the counterpart of the method and purpose of the natural sciences rather than an object of lived experience.
>
> (Kohák 1984: 12)

In the light of our lived experience, however, a commitment to materialism is difficult to sustain, for in the living of our lives we are presented with a world that is deeper, richer and in many respects stranger than materialists would have us believe. Pre-reflectively, the beechwood does not reveal itself as a collection of determinate material objects arrayed in Euclidean space. The place is suffused with meaning and value. In Roger Deakin's nice image (2008: xiii), one looks up at the canopy as if to the shallows from a seabed;

the darkness in the forest's depths is not merely an absence of light but a haven for hidden creatures; this or that branch discloses itself as one that could be grasped or thrown or snapped; that path reveals itself as the one leading to such and such a place. Space is not experienced as metres and centimetres, but as ground to be covered, as a felt sense of expansive freedom or a stifling sense of its lack. Time here is congealed in the trunks of trees, or it is failing sunlight filtered through their branches. It is the rise and fall of the sun and the moon, the turning of the seasons.

Just as phenomenology can 'awaken our experience of the world' (*PP*: 206), so environmental phenomenology can awaken our lived experience of the natural world. I shall argue that this is a good thing, ethically, aesthetically and (for want of a better word) spiritually. But it is also radical philosophy. First, phenomenological inquiries have the potential to undercut certain dogmas of environmental philosophy. So, for instance, in Chapter 2 I show that a prevalent form of scepticism regarding animal minds rests on an implausible account of how we relate to non-human animals in the living of our lives. In Chapter 3 I argue that the perennial debate about the metaethical source of natural values presupposes a false 'subjectivist' picture of the relation between human beings and the world. Furthermore, while these debates and problems are undercut, new avenues of philosophical inquiry are opened up. Environmental phenomenology is not just a new approach to the old issues. It is a way into a host of new issues as well.

Φ

I have been referring to 'nature' and what is 'natural'. But what exactly do I mean by these terms? The question must be asked since both words are used in a variety of ways. The natural is sometimes contrasted with the supernatural. And if one has no time for talk of gods, angels and demons, nature (the set of natural things) may be thought to encompass everything that exists. In other contexts, nature is opposed to culture (as in references to 'nature versus nurture'), yet in others it is taken to refer to the essential characteristics of a thing, such that individual things (and kinds of thing) are thought to *have* natures. To complicate matters further, 'nature' is often thought to have normative connotations – for instance, to say that an act is *against* nature is usually to say that it is morally wrong. The list could doubtless be extended. But there is little point in trying to provide a comprehensive account of all the ways the terms 'nature' and 'natural' have been used. And it would certainly be unwise to try to come up with a definitive, one-size-fits-all definition. Nonetheless since we need to have *some* idea of our subject matter, let me offer the following, very provisional definition:

A thing is natural to the extent that its current state is relatively unaffected by human agency.

On this definition, AstroTurf and human footprints come out as non-natural, since both are produced by humans (the fact that the footprints may not be *intentionally* produced is of no account). Birds' nests and beavers' dams, by contrast, count as natural (they're not the products of *human* agency). Hedgerows come out as natural, for although human beings originally created them, their *current* make-up typically owes more to natural forces (the weather, colonisation by wild plants, the local wildlife) than to human ones. For similar reasons, the Scottish Highlands may be considered natural because, although they were in large part formed through human action (clearances and the like), their current state is, by and large, the product of non-human forces. As Robert Macfarlane suggests, 'despite the human influences in their making' these places have 'become wild' – become natural, one might say (2007: 80).

Three points of clarification. First, I am using 'thing' in a very broad sense to denote, not just solid material objects but, more broadly, any part of the biosphere that can be named – so not just trees and pebbles, but clouds, and not just entities, however solid or insubstantial, but also substances (like fresh water) and even places, habitats and environments. Second, although I will discuss the various ways that our experience of nature is conditioned by social and cultural factors, I am not impressed by claims that nature (or the natural world – I will use the terms interchangeably) is nothing more than a social construction, or something of that order. (I explain why I think this sort of constructivism is false in Chapter 5.) Third, I allow for degrees of naturalness. So although I am willing to concede Bill McKibben's point (1990) that due to the influence of anthropogenic pollutants on the weather, no part of nature remains entirely independent of human influence, I would also want to say that the rainforest on the eastern slopes of the Andes is *more* natural than London's Hyde Park, not because the rainforest is not managed at all (it might well be), but because it is managed less intensively than the park.

In relation to the last of these points, it should be noted that allowing for degrees of naturalness does not invalidate all talk of what is natural. There are degrees of baldness, yet (unfortunately for men like me) it still makes sense to say that one man is bald while another is not. The same holds true of naturalness. Perhaps nothing on earth is entirely unaffected by the actions of Homo sapiens, yet some things are sufficiently uninfluenced to warrant the ascription 'natural'. There is no good reason to suppose that the only nature is pristine nature, untouched by human hand. Furthermore, when it comes to classifying something as natural (as being more natural than not) I am inclined to be accommodating. So, as I said, I am happy to affirm that hedgerows or the Scottish Highlands are natural. And for similar reasons I'm willing to call some gardens natural as well – particularly ones of a wilder sort.

It is not my intention, here, to trace out a conception of the natural that is in any way unfamiliar or controversial. On the contrary, my aim is to approximate the way the term is used in references to natural history. After all, one would not be surprised to tune into a TV programme called 'The Natural History of Britain' to find the presenter discussing badgers or old growth oak woodlands. One would not be surprised to find her saying something about hedgerows and gardens. She might even have something to say about ecological cycles or cosmic ones, especially their influences on life on earth. It is unlikely, however, that the presenter would devote much time to urban environments, save for the purpose of examining the non-human animals that live there or, perhaps, for drawing some analogy between human social behaviour and that of some non-human species.

Φ

The general aim of this book is to develop a phenomenology of the natural world, an environmental phenomenology. On the one hand, I develop a new phenomenologically based account of what the natural world is, one that does justice, or at least is meant to do justice, to its richness, ambiguity and depth. On the other, I set out an original phenomenologically based account of how we should (ethically, aesthetically and 'spiritually') relate to nature. My conclusion, in brief, is that when it is done well phenomenology necessarily involves the exercise of a particular (ethical, aesthetic and spiritual) virtue, which I call 'attention'. This counts as an *environmental* virtue because in modern capitalist societies it is readily developed and exercised with respect to specifically natural objects.

In making my case I focus on the tradition of existential phenomenology, as exemplified by the work of thinkers such as Heidegger, Sartre and Merleau-Ponty.[1] Heidegger and Merleau-Ponty are the main players, though other phenomenologists ('existential' and otherwise) play bit parts, as do a number of contemporary thinkers, such as Ted Toadvine and Iain Thomson. My primary aim is not the exegetical one of interpreting what the textbook list of phenomenologists had to say about the natural environment (with the exception of Heidegger, they didn't say a great deal). Instead, I shamelessly plunder their work in the hope of shedding light on the natural world and our relations to it. To make one point I might refer to Heidegger's work; to make another I might draw upon that of Merleau-Ponty.

This is a risky strategy, of course. Phenomenologists frequently disagree with one another, often markedly. One only has to compare Husserl's early conception of the phenomenological reduction with Heidegger's analysis of being-in-the-world, or the picture of human relations set out in *Being and Nothingness* with that defended by Levinas. The risk, then, is that in drawing upon the works of a number of phenomenologists I might produce a Frankenstein's monster of a thesis, formed of a hotchpotch of incongruous parts. And this would of course be no thesis at all, at least no *coherent* thesis.

But this danger can be averted. For one thing, on many points – and on most of the important ones – existential phenomenologists are in agreement. So, for example, one is justified in speaking of 'the phenomenologist's' rejection of Cartesian dualism because, despite their various disagreements, Heidegger, Merleau-Ponty, Sartre et al. are all of a piece in rejecting this conception of what it means to be human. When the differences between the views of different existential phenomenologists are very great, I sometimes opt for one view over another, calling the position I endorse 'the phenomenological one'. For instance, I believe that on the topic of embodiment, Merleau-Ponty's account is more illuminating than that of Heidegger. So in Chapter 1 I endorse a conception which is close to that of Merleau-Ponty on this issue, and refer to this position as 'the phenomenological view' on the understanding that this is not a view that would be accepted by all phenomenologists. In other cases, the approach I develop is of my own design and so genuinely mine, rather than someone else's.

Outline of chapters

The first aim of Chapter 1 is to introduce some prominent phenomenological themes – notably, that of being-in-the-world. Its second, more specific aim is to bring a phenomenological approach to bear upon the broad question of our place in nature. That question is often thought to imply a dilemma: either we are material through and through, material parts of a material nature, or mind-like part of us that is supernatural. But this dilemma is, I suggest, false. Adopting a phenomenological approach, I show that we may be considered parts of nature in at least two distinct senses, neither of which is either materialist or dualist.

I begin by arguing that we can be in nature in the sense of being *involved* with it. This is to say that the natural world can (and typically does) show up for us in the light of our lived concerns, that it matters to us. To make my case I refer to a specific kind of involvement – practical involvement – and in doing this I draw upon Heidegger's account of concern, as developed in *Being and Time*. I pay special attention to some of the questions raised by a particular kind of practical involvement, namely the phenomenon of being at home in a natural environment. Are antimodernist writers right to think that we are becoming an increasingly rootless species, alienated from the natural world? Can non-human animals be at home in an environment? And, if they can, do they inhabit their environments in the same way that we inhabit ours? Does the phenomenon of being at home in nature have any moral significance? Must someone who is at home in a particular natural environment be inclined to treat it well?

Having responded to these questions, I proceed to outline the second sense in which we may be considered parts of nature.

My contention here is that we *inhere* in nature, which is to say that we are in nature in a compositional sense, made of the same stuff as the things we encounter. To explain what this involves, and to distinguish my account from materialism, I refer to Merleau-Ponty's account of embodiment, as set out in *Phenomenology of Perception*. Having explained how, according to that account, our perception of the world is indelibly conditioned by our bodies, I move on to consider how our embodiment conditions our perception of specifically *natural* things.

In the final pages of the chapter, I argue that, given our involvement with and inherence in the world, a certain kind of scientific naturalism, and thus a certain kind of naturalistic approach to environmental philosophy, ought to be rejected.

Φ

I begin Chapter 2 by arguing that we can be regarded as parts of nature, not only in terms of *involvement* and *inherence*, but also in the sense that as subjects among non-human subjects we inhabit a multi-species *intersubjective* world. To justify this claim, I begin by examining Heidegger's conception of being-with (*Mitsein*), before moving on to investigate the work of Husserl and Merleau-Ponty, and what is, in my view, the basis for a more plausible and less anthropocentric account of intersubjectivity.

In the second section, I turn my attention to a question raised by this account of 'cross-species intersubjectivity', namely, what reasons do we have for thinking that non-human animals are conscious? In tackling this question, I begin by setting out the various ways a phenomenologist might respond to the sceptical claim that *no* non-human animals are conscious. I concede that none of these responses can *refute* scepticism regarding animal minds: despite the phenomenologist's best efforts, it remains logically (and perhaps metaphysically) possible that all non-human animals are bereft of phenomenal consciousness. However, I contend that a phenomenological approach can *undercut* this kind of scepticism by rendering problematic the general conception of interpersonal relations on which it rests.

Having addressed the problem of scepticism, I turn to the question of how one can determine whether particular animals (and by implication, particular kinds of animal) are conscious. Here

I examine some of the methods employed in the study of animal behaviour, particularly those of cognitive ethology. The first of my two conclusions is that if one is to discover whether a particular animal is conscious it will, in many cases, be unwise to proceed on the assumption that the creature must, so to speak, prove itself through its behaviour to be conscious. My second conclusion is that ethological methods are often interpretative rather than inferential, which is to say that if a study yields a positive result, a particular behaviour will typically have come to disclose itself to the investigator, not as *evidence* of conscious, but as a *conscious behaviour* – as an angry swish of the tail, perhaps, or a contented purr.

In the third and final section I consider how in the light of this account we are to understand those cases when ethological investigations make no headway. In some cases, I suggest, we are faced with a failure of interpretation: the animal's life is mysterious not because it is the outward sign of an unobservable mental cause, but because we cannot decipher it. In other cases, however, the consciousness of the animal is called into question. In such cases, it might seem that we are presented with two options: either the animal in question is conscious in much the same way that we are conscious, or else it is entirely non-conscious, like a rock. Drawing upon Heidegger's work and that of Merleau-Ponty, I argue that this dilemma is in fact false, and that matters are more complicated, and more interesting, than is conventionally supposed.

Φ

In Chapter 3 I turn to the question of how we are to understand the various ways in which nature matters to us in the living of our lives. Taking my cue from Heidegger's account of involvement (*Befindlichkeit*), I suggest that something can matter to us without our valuing it, and to support this contention I examine the testimonies of a selection of unfortunate individuals who hate or are indifferent towards certain parts of the natural world.

My main concern in this chapter is, however, with the notion of an environmental *value*. In examining this notion, I begin with Heidegger's claims that 'values-thinking' is inherently 'technological', a manifestation of a domineering and exploitative comportment towards the world, and the natural world in particular. I suggest that these bold claims cannot be justified. Against Heidegger, there is nothing *inherently* pernicious about talk of value. Nonetheless, I contend that there is something problematic about the notion that *all* the many and various ways in which nature matters to us – and, moreover, all the ways that it matters to us *morally* – can be cashed out in terms of environmental values. In support of this claim I refer to Christine Swanton's criticisms of 'value-centred monism', as set out in *Virtue Ethics: A Pluralistic View*, and to Alan Holland's recent argument that environmental ethicists ought to be more concerned with the preservation of meaning than with the promotion of value. I conclude by tracing out the contours of a radical moral pluralism, radical in that it postulates, not merely a host of incommensurable values, but a variety of different morally relevant factors, including different sorts of bond, need and meaning.

Φ

Chapter 4 is concerned with moral normativity, the question of how we should act with respect to the natural world. I begin with the allegation that phenomenology is merely a descriptive exercise which might be able to open our eyes to the manifold dimensions of our moral lives but which will not be able to recommend any particular course of action. In response, I draw upon the work of Iris Murdoch to argue that expertise in phenomenology involves the development and exercise of a particular moral virtue, namely *attention*. Hence a training in phenomenology is to some extent a moral training. Or, to be more precise, attention is a moral virtue, not only for eudaemonistic reasons, not only, that is, because it benefits the attentive agent, but for what might be called world-directed reasons as well. It is, I propose, a virtue because it enables us to 'see' things fairly or justly, as they really are.

Having set out my arguments for this conclusion, I turn to the question of how this bears upon the natural world. My argument here is based on the general claim that the development and exercise of attention partly depends on one's material circumstances. Mass-produced objects, I suggest, tend not to invite attention. By contrast, natural things do. Hence, in modern capitalist societies, the presence of nature tends to foster the development and exercise of attention. In conclusion, I argue that we ought to conserve nature in two senses. On the one hand, we ought to conserve physical nature: we should take steps to curb pollution, to protect natural habitats, to conserve endangered species and so forth. But on the other hand we should try to conserve the meaning nature has for us as a realm independent of our practical concerns. In Heidegger's apt phrase, we should let it be.

Φ

In the fifth and final chapter I consider what my phenomenological approach (or indeed any such approach) might be able to say about the notion that the natural world exists in itself, independently of human concerns. Many phenomenologists would regard any such notion as incoherent. Yet such a stance – ostensibly one of metaphysical antirealism – is unlikely to be welcomed by environmental thinkers. For if nature is to command our respect, then it might seem necessary that it present itself to us as existing independently of our 'merely human' concerns.

The charge, then, is that the phenomenological position on these matters is both anti-realist and anthropocentric, and hence at odds with an attitude of respect for nature. I respond to it in three ways. First, I concede that, according to the phenomenologist, no sense can be made of a world that is not 'lit up' in terms of our interests, practical concerns and so forth. However, I contend that this thesis does not amount to anthropocentrism, since, in this context, the set of interests, practical concerns, etc. that we designate as *ours*, includes the interests, practical concerns, etc. of at least some non-human animals.

Second, I accept that phenomenology cannot accommodate the notion of a world that is radically independent of our concerns, but I suggest that this is no cause for regret, since any such notion is incoherent. Indeed, once one relinquishes the notion that there is a way the world is 'in itself', it becomes clear that phenomenological investigations are especially well suited to elucidating the various ways in which nature discloses itself to us as an independent reality. To illustrate this, I examine (1) the disclosure of the world as an object, not just for us, but for non-human subjects as well and (2) its disclosure as a realm indifferent to (and in this sense independent of) human concerns.

In developing my third response to the charge of anthropocentrism, I begin by investigating the claim that, although whatever presents itself to us in experience necessarily does so in the light of our concerns, the 'process' by which anything presents itself at all is partly transhuman – an event of Being, perhaps, or an intertwining of a 'flesh' that 'traverses' us. I argue that, heard in the right way, claims of this sort can awaken us to the presence of that more-than-human 'nature' which moves within us when we perceive anything.

Φ

I hope these arguments will demonstrate the benefits of a phenomenological approach to environmental philosophy. But as they say, the proof of the pudding is in the tasting. So with this in mind, it is time to turn to our first topic, the question of our place in nature.

Note

1. On the different kinds of phenomenology, see Moran 1999.

1
Our Place in Nature

1.1 Being-in-the-world

As we saw in the Introduction, the term 'nature' has several meanings. According to a long-standing tradition, it denotes the part of reality that is material rather than soul- or mind-like, *res extensa* as opposed to *res cogitans*. On this dualistic conception, human beings are thought to be partly natural (on account of their having bodies) and partly non-natural (on account of their having souls or minds). So although a man has a certain size, shape and mass, he is thought also to possess a radically non-natural part, a soul or a mind.

Environmental thinkers typically reject claims of this kind. The thought that human beings are essentially *other* than nature can, they contend, all too easily transform into the thought that human beings are essentially *better* than 'merely natural' things. And this hierarchy, when supplemented by certain assumptions about the moral permissibility of dominating 'lower-order' beings, can all too readily serve to legitimise the domination of nature by humans (see further, Warren 2001: 190). In this way, the abstract metaphysical claims of dualists such as Descartes are said to have prepared the way for the exploitation of nature, and hence for the host of environmental problems that face us today.

So what are the alternatives? We live in materialist times, and one popular option is to contend, against the dualists, that humans are essentially natural beings in that they are essentially material beings (the question of what *matter* is, is often left unanswered). This is not, however, the line I will take in this chapter. In the following pages,

I will articulate and defend an account of our place in nature that differs, not only from materialism but from any kind of ontological monism, and not only from monism but from ontological dualism as well.

Φ

My approach will be 'phenomenological'. But what exactly does this mean? According to a prominent conception, epitomised by the works of Merleau-Ponty, the primary aim of phenomenology is to describe, in as unprejudiced way as possible, how we find ourselves in the world. For a description to count as unprejudiced in this sense it must, among other things, be free of prior theoretical commitments about what reality is like. Instead the phenomenologist – or at least the good phenomenologist – approaches her experience with an open mind, leaving her theoretical commitments (to materialism or idealism or whatever) at the door.

At first sight, these aims might seem very modest. After all, to describe how I find myself is, one might think, to describe the content of my experience, and the content of my experience must be immediately evident to me. Before me, at this very moment, I can see a table set for dinner – a fork, a knife and a spoon, neatly arranged on a red tablecloth. The atmosphere in the restaurant is hot and humid, although there is a slight breath of air from a fan above my table. It turns with a regular click, click, click like a horse trotting on a paved road. Outside, the traffic on the beach road is light – just the occasional motorbike or auto-rickshaw. Beyond the road, there is nothing but hot dark space. I can hear the airy rumble of waves on the shore. I can feel the brooding presence of the ocean.

One might think that all this is in itself of no philosophical significance whatsoever. It is obvious that these things *seem* to surround me. It is clear, for instance, that I appear to be surrounded by a collection of material objects. Surely the task of philosophy is to investigate the relation between these appearances and reality, to determine whether things really are as they appear.

For the phenomenologist, however, there is nothing obvious about how things present themselves to us in experience. Quite the opposite: Husserl and his successors stress that an adequate phenomenological description must come not at the beginning of one's inquiry but at the end, as an accomplishment, the fruit of rigorous and disciplined attention and reflection.

So with this in mind, I return to my experience, not only to *what* is experienced, but also to *how* it is experienced. The first thing I note is that I am not primarily in the world as a spectator. As I ponder these matters I am already interacting with a world that outstrips my consciousness of it. I am sitting on a chair: its solid back is pressing slightly against my spine; the wooden arm is hard beneath my right elbow. I'm taking all this down on a grey, spiral-spined pad from Woolworths with a pen a removal man left in my flat back in the UK. I use the pen without reflecting – smoothly, efficiently, as if it were an extension of my hand. The fork, knife and spoon disclose themselves as being within my reach, the neighbouring table as being a few steps away. As Sartre would say, the locations of these items are not 'defined by pure spatial co-ordinates but in relation to axes of practical reference' (*BN*: 321).

In describing how things are for me, I do not first describe how I am and then a world I apprehend. On the contrary, to describe how I am is to describe a world I inhabit. What is more, it is a world I find myself always already inhabiting; interacting with things (for instance) even before I step back to reflect on that interaction. So to specify how I am at this precise moment, I find that I must refer, not just to the various things with which I am pre-reflectively interacting (the chair, the pen, etc.) but also to the general situation in which I find myself (sitting in a restaurant) and the various possibilities that that situation affords (primarily, the possibility of dinner). And in all this, I find that I must refer to *others* – the waiters chatting by the stairs, for instance, or the potential presence of other diners. True, I can conceive of myself as an isolated subject confronted by an objective order of things. But that is not what presents itself to me prior to reflection. Pre-reflectively, I am presented with what Heidegger calls a unitary complex of being-in-the-world.

Furthermore, if I find myself already in-the-world in this sense, unreflectively engaged with it, then the world for its part bears the stamp of my being, which among other things is to say that it reflects my concerns – my projects, purposes, practical commitments and the emotions to which they are connected. So the pen reveals itself as a thing because of the place it occupies in a field of practical significance, one focused, at this precise moment, on my current project of writing these words. It feels comfortable, reassuring between my fingers (in fact, I notice that it has 'Comfort Grip' written on the side). In general terms, it shows up in my experience as a distinct thing because it plays a role in my life. The atmosphere is humid, oppressive; yet this is not a mind-independent state of affairs, but rather a function of my northern European understanding of the world. The truth is that just as my being is worldly, so is the world 'human', traced out in terms of my lived concerns and, more generally, those of beings like me – and, it may be added, the manifold cultural, social and historical conditions those concerns reflect.

Φ

I will not say any more right now about what it means to be 'in-the-world' (though I will have much more to say on the topic later on). Nor will I spend a great deal of time defending the phenomenologist's account (though I will respond to a number of objections that can be raised against it in subsequent chapters). Instead, I will move straight on to consider how the phenomenological account of being-in-the-world may be brought to bear upon our experience of natural environments. And in particular I will offer some preliminary thoughts on how it might be applied to the question of our place in nature.

At first blush, the phenomenological account just sketched might seem very much in tune with a certain kind of approach to environmental philosophy. For, whatever else it is, it is clearly opposed to the dualistic claims canvassed in the opening paragraph of this chapter. For existential phenomenologists such as Heidegger, we are *in* the world, not ontologically estranged from it, and this might be regarded as a salutary conclusion, morally speaking. After all, if, as many environmental thinkers claim, it is a bad thing to be a Cartesian dualist, then the essentially non-dualistic nature of the phenomenologist's account must count in its favour. What is more, many environmental thinkers will applaud the phenomenologist's attempts to do justice to our emotional and practical engagement with nature. For on the phenomenological account, the view that nature is a disenchanted realm of neutral stuff represents, not how things really are, but rather a high-level abstraction, arrived at through a kind of falsification of the world we experience.[1]

To discover what nature really is, the phenomenologist maintains, we must set aside the abstractions of science and scientistic metaphysics to examine the nature we encounter in the living of our lives. And *this* nature, the nature we inhabit, is shot through with our distinctively human concerns. As a Heideggerian phenomenologist might say, our being-in-nature is necessarily marked by 'care' (*Sorge*).

But one must take care not to conclude too much from all this. For one thing, although environmental thinkers are pretty much unanimous in their rejection of 'dualism', they do not always make it clear what position they are rejecting. As Baker and Morris (1995) have shown, attacks on 'Cartesian dualism' are often levelled, not at the position Descartes himself advocated, but at a caricature of that position and a 'strawman' Descartes. Moreover, even if one brackets the question of whether the dualistic thesis set out above can be rightfully attributed to Descartes, the fact remains that one can reject that thesis and yet, with perfect consistency, consider all non-human beings unworthy of moral regard. So, for instance, consider Oswald Spengler's view, as set out in *Man and Technics*, that the human is in essence a 'beast of prey' subject to 'carnivore' ethics, and hence 'obliged to maintain itself against Nature and to give its own being some sort of significance ... some sort of superiority' (1960: 9, 19, 26). Or consider Baruch Spinoza. Despite his conviction that humans, and indeed all things, are parts of a single reality, 'God or Nature', Spinoza himself was an inveterate anthropocentrist, for whom 'the lower animals' may justifiably be used 'at our pleasure' and treated 'as is most convenient for us' (1996: 135).[2] Furthermore, if rejecting ontological dualism is not sufficient to inspire environment friendly behaviour, it is not necessary either. There is no reason why a proponent of ontological dualism must be motivated to act badly in her relations with natural things. Christian thinkers are presumably ontological dualists of one kind or another, yet a significant number of them maintain that our moral duty is to care for the natural world, to act as responsible stewards of God's creation.

So it would be hasty to conclude that dualism must be a Bad Thing, environmentally speaking, and that it must therefore be a Good Thing, environmentally speaking, to reject it. And there are also good reasons to doubt the truth of the crude claim that phenomenology enjoins us to care for natural things. It is true that Heidegger speaks of our being-in-the-world in terms of 'care', yet he does not present this as a normative claim. He is not saying that we *ought* to care for things. Instead, he is suggesting that to be in-the-world is necessarily to care. He is making the entirely general and non-normative point that we live our lives, not as dispassionate observers, but as actors who are always already engaged with the world. Furthermore, this is not to suggest that our relations with natural things are always warm and fuzzy. For one thing, 'care' may be exhibited in natural or urban environments alike. For another, the specific form of this engagement remains unspecified. Feelings of revulsion, hatred or indifference towards natural things are, on Heidegger's account, all manifestations of care (see further, Thomson 2004: 389). To be nauseated by natural things, like Sartre's Roquentin, is still to be in-the-world.

1.2 Involvement

Before considering how exactly the phenomenological account of being-in-the-world can be applied to the question of our place in nature, we need to distinguish two interrelated senses in which we are in-the-world. First, we are *involved* with it, which is to say that whatever shows up in our experience does so because it connects with our lived concerns in some way, because it matters to us. Second, we *inhere* in it, which (very roughly) is to say that we are made of the same stuff as the worldly things we encounter.

I do not claim that this analysis is exhaustive. We are in-the-world in other ways as well. (Indeed we will examine one such way in the next chapter.) But in the following pages I will try to show that, however else we are in-the-world, we necessarily find ourselves involved with a world in which we inhere. I will try to show that our being-in-the-world is characterised both by involvement[3] and by inherence.

Φ

Let's begin with the first of these two 'i's. In the simplest terms, to be involved with a world is to find oneself in a context that speaks to one's concerns. But when one probes deeper to ask what exactly this might mean, things quickly become much more complicated. For only a little reflection reveals that each of us is involved with the world in a wide variety of ways, depending on whether we are making lunch, for example, or walking to work or chatting with friends. Different people, moreover, are involved with the world in different ways. The world speaks to the concerns of a middle-aged mother and to those of her teenage daughter, but it speaks in very different ways. And (to continue the metaphor) it will say different things to a mother and daughter from another culture. In sum, our involvement with the world will reflect a host of interrelated factors, including our personal projects, personalities, social roles and cultural backgrounds.

And all this holds true of the *natural* world as well. Reflecting phenomenologically, it quickly becomes clear that humans can be involved with the natural world in a multitude of ways. Consider an area of wetland on the outskirts of a town, for example. It will speak to the concerns of a number of individuals, representing a variety of social and cultural conditions. It will speak to the hunter in one way, but to the birdwatcher in another. And it will speak to the poet, the painter, the town planner and to a whole host of others too. Each of these individuals will be involved with the area in one way or another, yet their specific modes of involvement will be quite different.

Φ

Many of these modes of involvement can be understood in terms of affect, which is to say that to be involved with a natural environment is often to *feel* a certain way about it. We will consider this sort of involvement in Chapter 3. In this section I focus on another broad way in which we can be involved with the natural world, namely *practical* involvement. (Like Heidegger, I concentrate on this kind of involvement, not because the world can only show up for us in the light of our practical dealings, but because focusing on our practical involvement is a particularly good way of explaining what it means to be involved in a world (cf. Foltz 1995: 34, n.3)).

The first thing to note about this practical involvement is that, for the most part, we do not notice that it is going on. Reflecting on Descartes' *Meditations*, I might find myself wondering whether the external world exists. What I fail to notice, however, is that while I am pondering these matters I am unreflectively engaging with a world – leaning back in my chair, perhaps, or twirling my pen.

I am always already interacting with things, secretly in communion with the world, even while at the level of reflection I question its existence.[4] And of course this doesn't just apply to philosophers: each of us spends a large part of our lives unreflectively engaging with a range of things – sitting, standing and leaning on them; reaching for them or finding them just out of reach; peeling them and slicing them; sucking them and biting them; hiding behind them; throwing them and catching them – and so on. Indeed most of the things we encounter in the world disclose themselves to us in the context of this unreflective practical engagement: as Heidegger said, they disclose themselves as 'ready-to-hand' (*zuhanden*).

When any particular thing discloses itself in this way it does so as a nexus in a network of meaningful relations. So, for example, the pen I am twirling between my fingers refers to the presence of various other items in the world – the paper on which I am writing, the desk at which I am sitting, the chapter I hope to finish and so on. Ultimately, this 'referential totality' must be understood with reference to the concerns of a particular individual. In the present example, the whole context of pens, paper, desk, etc. must be understood with reference to my desire to write a book. For someone with different concerns a different world would present itself. So a child, for instance, might see my office as a world of possible playthings; a cleaner might see a world of things to be tidied up, dusted and polished; a health and safety officer might see a collection of hazards. All three might be familiar with the same volume of physical space, but they inhabit different (yet no doubt overlapping) worlds.

I am able, for the most part, to operate fluidly and unreflectively in my office-world. Yet our practical interactions with things are not always fluid and unreflective. Things buckle, break, snap and split. When some problem or fault prevents my unreflective engagement with the thing, it discloses itself as what Heidegger calls a thing 'present-at-hand' (*vorhanden*). So if in bringing my pen to paper I were to discover that its nib was broken, my fluid engagement with the world would then have been disrupted. The pen would have stood out from the background of the world, wrenched from the context of relations within which, in my pre-reflective dealings, it disclosed itself as ready-to-hand. It would have revealed itself to me as obtrusive, standing out from the referential totality in which it is usually immersed.

Φ

This account of practical engagement illuminates one way in which we can be involved with natural environments: One is *at home* in a natural environment if it makes sense; if one is able, smoothly and for the most part unreflectively, to find one's way about in it.

Two things need to be noted here. First, one's being at home in a particular natural environment in this sense need not be accompanied by a specific affective tone. The hunter and the naturalist may both be at home in a particular forest, yet their *feelings* about the forest might be very different.

The second thing to note is that we are referring to *particular* natural environments. We are not dealing with the global environment ('The Environment'), but with an older meaning of the term: environment as milieu or *Umwelt* (see further, Cooper 1992). Indeed it is not clear what it might mean to be at home in The Environment or in The Natural World *per se*. One can, with difficulty, imagine someone who is at home in any natural context – some particularly seasoned explorer, perhaps. But someone who is at home in The Environment? That doesn't make sense. The Environment is simply too *big*.

Φ

To be practically involved with a certain natural environment can, therefore, amount to one's being at home in it. To convey what exactly it means to be at home in this sense, it will be useful to focus on a concrete example. But where to find one? Some may think it unlikely that we'll find any examples close to home. After all, aren't we rootless moderns becoming less and less at home in natural contexts? And so surely we would do better to look outside the post-industrial West, either backwards in time to some pre-modern age, or sideways in space to some pre-modern but extant culture?

These anti-modernist views are difficult to appraise. If the claim that we are becoming increasingly alienated from nature is taken to be an empirical hypothesis, then it is one that is imprecisely formulated and difficult to test. Nonetheless, there are some reasons (though perhaps not compelling ones) for thinking that it is true. For one thing, there is, in some parts of the world at least, a drift towards *urbanisation*, which, one presumes, would mean that less and less people in those areas are able to develop an unreflective familiarity with natural things. Then there is the increasing *mobility* of individuals. Nowadays, many of us do not stay in one place long enough to build up a rapport with our surroundings. As Richard Mabey, for one, has noted, the freedom to choose where to live brings with it the danger of estrangement from nature (2005: 74). Third, there is the influence of *technology*. Even the most ardent technophiles would have to concede that modern technology often serves to alienate us from nature. Electric light, heating and air-conditioning obscure the annual and daily rhythms of the natural world. We enjoy heat in winter and cold in summer, foods out of season, nights bathed in fluorescent light. Perhaps office-bound, technologically savvy people have developed other practical competencies – perhaps, for instance, they have learnt to be at home on the Internet – but they are not at home in nature.

I suspect that much of this is true. However, one cannot avoid the fact that these are in part empirical claims and, as such, should not be accepted without empirical evidence. It might *seem* that we are becoming increasingly alienated from nature, but is this really the case? What then are we to make of the current upsurge of interest among suburbanite Britons in gardening, birdwatching and tending allotments? Is this evidence that, contrary to the gloomy assessments of romantic anti-modernists, a significant proportion of us are still at home in natural contexts – not in wild places, but in suburban gardens and meadows on the edges of towns? These are at least in part empirical questions, and as such not best tackled by philosophers.[5]

Nonetheless, regardless of whether we are becoming increasingly alienated from the natural world, it remains the case that many of the best examples of what it might mean to be at home in nature have a distinctly counter-modern, romantic tenor. Indeed, for a good example of what it might mean to be at home in a natural context, one could do no better than to look to the writings of nineteenth-century nature lovers – men like William Wordsworth or Thomas Hardy. So, for instance, one could look to the character of Michael, in Wordsworth's poem of that name: an old shepherd, for whom the fields and hills had, 'like a book, preserved the memory / Of the dumb animals, whom he had saved / Had fed or sheltered' (1994: 120). Or one could consider the character of Gabriel Oak from Hardy's *Far From the Madding Crowd*, a shepherd, like Michael, who knows his flock and is able to read the land.

Michael and Gabriel are both at home in natural environments, one on 'the forest-side in Grasmere Vale', the other in the softer downlands of Hardy's Wessex. This is particularly evident in their ability to read the weather. So we are told that Michael, being 'watchful more than ordinary men', 'had learned the meaning of all winds', and that 'oftentimes, / When others heeded not,' he found himself able to make out the distant 'subterraneous music' that betokens a storm (1994: 119–20). The following passage, related from Gabriel's point of view, suggests that Hardy's shepherd has a similar ability:

> The night had a sinister aspect. A heated breeze from the south slowly fanned the summits of lofty objects, and in the sky dashes of buoyant cloud were sailing in a course at right angles to that of another stratum, neither of them in the direction of the breeze below. The moon, as seen through these films, had a lurid metallic look. The fields were sallow with the impure light, and all were tinged in monochrome, as if beheld through stained glass. The same evening the sheep had trailed homeward head to tail, the behaviour of the rooks had been confused, and the horses had moved with timidity and caution.
>
> Thunder was imminent, and, taking some secondary appearances into consideration, it was likely to be followed by one of the lengthened rains which mark the close of dry weather for the season.
>
> (1978: 297)

Michael and Gabriel Oak are both able to read the skies. To be sure, some of this knowledge will be theoretical or propositional, of a kind that could be recorded in a book. So Michael might know where a certain kind of bird likes to nest, or what type of weather favours the flourishing of a particular species of plant, and these items of knowledge could be set down in a book on natural history. Moreover, some of the practical conclusions arrived at will be the result of inference. The shepherd feels a certain kind of breeze from a certain direction and then infers that a storm must be approaching from the southwest. But in other cases the meaning of the signs will be directly perceived, no inference being required. Imagine Michael standing on some hillside. The breeze discloses itself to him, not as an object present-at-hand to which a meaning is subsequently attached, but as ready-to-hand, as already meaningful.[6] It means a storm is coming. And it discloses itself to Michael as meaningful because of the way it refers to other things that matter to him – the river in the valley (already high, he thinks); the lack of cover on this exposed patch of land; the fates of his sheep, and with them, his fate. It is this non-inferential rapport with things that marks the shepherd out as being at home in nature, in the sense we are considering.

Heidegger calls the faculty by means of which things ready-to-hand are identified, circumspection (*Umsicht*), and this, he says, is a kind of sight (*BT*: 216). But, as he himself would admit, to be at home in a certain environment is not merely to *see* it in a particular way. On the contrary, in unreflectively interacting with natural things a number of senses are typically brought into play. For Michael, the 'subterraneous music' that heralds an approaching storm discloses itself as ready-to-hand, yet the medium of circumspection here is sound. At other times, he might catch the scent of a particular species of plant, or he might chew on a leaf to determine whether it will make good eating for his flock. He might feel a certain kind of breeze, or more generally, 'something in the air', a change in pressure, perhaps. He might notice that the ground has a distinctive consistency; that the grass is young and tender or old and tough. He is quite literally in touch with the natural world.

Φ

It is not only human beings that can develop such a rapport. Animals – sentient ones, at least – are typically *au fait* with their surroundings; indeed they tend to be *more* at home in their surroundings than humans, in part, no doubt, because they lack the capacity, through reflection, to step back from the contexts in which they find themselves. True, the experiences of animals are in many respects unlike those of humans, yet for all these differences, it seems likely that things disclose themselves to at least some animals as meaningful, and it also seems likely that they do so because of the ways they indicate other things. As David E. Cooper points out, the referential totalities of which Heidegger writes would seem to obtain in non-human contexts: '[a]nimals, too, dwell in fields of significance; the droppings at the entrance of the tunnel indicate a fox, which signifies a threat to the badger's young, whose squealing expresses hunger, which refers the badger to the berries behind that tree, the scent on which means the recent presence of a fox, which indicates ... etc.' (1992: 170).[7]

To be human (or indeed to be a badger) is to be surrounded by a penumbra of items from which, at a pre-reflective level, one does not distinguish oneself. To learn how to function in a particular context involves an extension of this penumbra (which is not to say that it *only* involves this). Items that were once present-at-hand become ready-to-hand. After a while, the clutch and the gear-stick, which formerly took so much conscious thought to manipulate, become less obtrusive and eventually melt entirely into the unnoticed penumbra of circumspective concern that constitutes one's unreflective being-in-the-world (cf. *PP*: 143).

And this also holds true of our interactions with natural things. In learning how to track an animal, for instance, certain items (those strangely shaped marks in the snow, that odd, peculiarly pungent whiff) start to make sense (fox tracks! fox urine!), and after a while one finds that one is engaging with these 'things' in much the same fluid, unreflective way that one engages with familiar household objects. In this way, the penumbra of a master tracker expands to incorporate all manner of things which, to more urban folks, would either be misinterpreted or overlooked entirely.

This transformation could be interpreted in an egoistic and anthropocentric way. One could say that the tracker's 'penumbra' has expanded so as to include more and more of the natural environment, that he has made the environment his own (cf. Cooper 1992: 178). But the process of becoming at home in a natural environment need not be understood in such terms. In *Crow Country*, Mark Cocker describes his increasing sense of familiarity with a particular stretch of woodland, 'Lightwood'. Although he came to feel that Lightwood was his, he didn't feel that he owned it, but rather that he 'belonged completely to it' (2008: 17). His point, surely, is not that he appropriated Lightwood in circumspective understanding, but that, if anything, Lightwood appropriated him.

This is a union with nature of sorts. But it does not involve one's absorption into some kind of monistic whole, 'Nature' with a capital 'N'. Nor does it involve one's transpersonal identification with a more-than-human reality (against Fox 1995). It is better conceived in terms of harmony. To learn to be at home in a particular natural environment is, in part at least, to attune oneself to it, not necessarily through the quiet contemplation of natural processes, but by learning to find one's way about the place in a good solid dirt-under-the-fingernails kind of way: learning where to tread, what to eat, how to look and listen and so forth.

Φ

Let us pause for a moment to consider the moral implications of all this. At first sight, it might seem that someone who is at home in a particular natural environment must necessarily be inclined to value that environment. After all (one might think) it can't be a bad thing to treat one's environment as if it were one's home. For to treat it as one's home must be to value it as if it were one's home. And who doesn't value their home?

But matters are more complicated than they might at first appear. For one thing, some people are quite at home in natural environments about which they couldn't give a damn. A greedy developer, for example, might be entirely *au fait* with a stretch of marshland he nonetheless intends to drain. And others might value a natural environment in which they feel at home, not because they value that environment in itself, but because it allows them to achieve some other end. For instance, a big game hunter at home in the savannah might support efforts to conserve that kind of habitat, but only, ultimately, because he wants to add more heads and pelts to his trophy collection.

So being at home in a particular natural environment is not sufficient to generate a positive moral concern for that environment. Nor is it necessary, which is to say that in order to value an environment one does not have to be at home in it. Most people who care about protecting the rainforests of South-east Asia wouldn't know a chevrotain from a Chevrolet.

Being at home in a particular natural environment is therefore neither necessary nor sufficient to generate a positive moral regard for that environment. But this need not be regarded as a cause for too much regret. For there is no need for us to fall prey to modal anxiety here, no need, that is, to be held hostage to the stringent demands for necessity and sufficiency. For being at home in some natural environment might *tend* to foster moral concern for that environment, even if it doesn't necessarily do so and even if it is not enough, in itself, to generate such concern.

Empirical studies drawn from the field of environmental education suggest that this is in fact the case (Palmer et al. 1999; Corcoran 1999). In a series of studies, environmental educators were asked to write autobiographical statements describing the formative influences that led to their concern for the environment. For sample populations in Australia, Canada, South Africa and the UK (but not, interestingly, for those in Hong Kong, Sri Lanka and Uganda), childhood experiences of nature were found to be the most significant such influences (Palmer et al. 1999: 454).

Many of these childhood memories were of being at home in nature, in the sense that we have been considering. Thus for one participant, as for many others:

> the outdoor world all around me as a child was secure and alluring. The back border of our property was a brook where I spent countless hours. From its beginning in a small marsh at the top of the hill, to its disappearance under a culvert at the bottom, I knew its every rock, whirlpool and miniature waterfall.
>
> (Corcoran 1999: 212)

The conclusion drawn by the researchers is clear:

> the results presented here suggest the importance of providing young people – indeed people of all ages – with opportunities for positive experiences of nature and the countryside. It is those 'in' and 'with' the environment experiences that appear to be fundamental to the development of long-term environmental awareness and concern.
>
> (Palmer et al. 1999: 199)

The implication is that childhood experiences of being at home in nature might tend to foster environmental concern, even if they are not sufficient to generate such concern and even if such concern can be produced in other ways.

1.3   Inherence

We have been considering our place in nature in terms of practical *involvement*. To be in-the-world is, we have seen, to be practically involved with some sort of world or environment. In natural contexts, it is to be so involved with a natural environment – to be at home in it, for instance. But to be in-the-world is not just to be involved with the world in this way but also to *inhere* in it, to be a part of it in a compositional sense. Merleau-Ponty, for instance, maintains that we are 'of the same stuff' as the world (*N*: 218), 'caught' in its 'fabric', our 'cohesion that of a thing', albeit one that 'holds things in a circle around itself' (*PrP*: 163). To understand what he means, we need to return, as ever, to our lived experience; and in order to do this, it will help to begin with a particular phenomenon.[8] So: right at this moment, there is before me a table covered with a blue and white checked tablecloth, and on the table, just within reach of my left hand, a book – appropriately enough, *My First Summer in the Sierra*, by John Muir. I want to say that I perceive *the book*, rather than merely an appearance of it. Yet when I attend closely to my experience I note that I am presented with only three of its six sides, the front cover (a photograph of the famous Yosemite Dome), the side opposite the spine, and a sliver of the bottom side. I focus on the upper side, the book's cover. I know that it is rectangular, yet at this angle and because of the way it is curled upwards, it seems a different shape, an irregular blue and green quadrilateral. The shape doesn't seem to vary when I move my eyes, but it does shift in and out of focus. As I rock my body backwards and forwards and to the side the shape changes. The further I lean forwards, the more rectangular the cover appears.

My perception of an irregular patch of blue and green is, one might think, merely an appearance, where this 'merely' implies that there is something lacking in my perception, that this isolated visual impression can capture only part of what the book is. Yet this is a conclusion arrived at through reflection. Pre-reflectively, I do not seem to be perceiving merely an appearance of the book, one that captures only a fraction of what the book is 'in itself'. On the contrary, and as I said, I want to say that I am perceiving *the book* (cf. *PP*: 324, *CES*: 157–8).

How can this be explained? Merleau-Ponty's explanation (and I concede that there are others) appeals to the role of the body in perception. He begins by noting that our bodies dictate optimum conditions for the perception of different kinds of object. As Taylor Carman explains, in Merleau-Ponty's view 'we have – and *know* and *feel* ourselves to have – *optimal* bodily attitudes that afford us a "*best* grip" on things, for example the right ... angle from which to see something' (2008: 110). So when I perceive my copy of Muir's *First Summer* lying at an oblique position in relation to me, I feel the obliqueness of the book's position as a 'lack of balance' (*PP*: 302). If my bodily attitude towards it were *optimal*, if, for instance, I were holding the book before me, as if to read its title, its cover would present itself to me as rectangular. But as it lies before me on the table, one edge presents itself as being further from me, my bodily attitude does not afford me optimal grip on the object, and as a result it discloses itself to me as unbalanced. The upshot of this is that I do not find myself presented with one appearance of the book (the book from angle *a*), which would have to be synthesised with other appearances (the book from angles *b*, *c*, *d*, etc.) to form a conception of the book 'in itself' (cf. *PP*: 324). No, I perceive *the book itself* in an 'unbalanced', oblique position.

Distance can be treated in a similar fashion. Were it very far away, the book would appear to be tiny. Yet this visual impression would not constitute one piece of 'raw data', which would have to be synthesised with others in order to determine the book's actual size (cf. *PP*: 300). The book would disclose itself as being *too* far away, as in other circumstances I would perceive it as being *too* close. In either case, my distance from the object is, as Merleau-Ponty puts it, 'a tension which fluctuates around a norm' (*PP*: 302), where both the tension and the norm can only be understood with reference to certain optimal bodily attitudes. In the most general terms, 'any perception of a thing, a shape or a size as real ... refers back to the positing of a world and of a system of experience in which my body is inescapably linked with phenomena' (*PP*: 303–4). To inhabit a world is therefore to *inhere* as a corporeal being in a world with which one is also *involved*. The two moments of being-in-the-world are inextricably related.

In sum, Merleau-Ponty's conclusion is that 'the system of experience is not arrayed before me as if I were God, it is lived by me from a certain point of view' (*PP*: 304). And this point of view is that of my body. But this does not constitute a limitation; one should not conclude that it would be *better* if I could bypass the appearances and adopt a God's eye view of the object as it is in itself, from all angles at once. For such a state of affairs could not be the object of any possible perception; hence it would be no thing at all (cf. *BN*: 306–7). We should not lament the fact that we are unable to attain such a view from nowhere any more than we should regret not being able to experience morning, afternoon and evening all at once.

Φ

On this Merleau-Pontian account we are therefore corporeal or embodied. This is not to say that we just so happen to be corporeal, as if we might have lived our lives as Platonic souls coursing through the Realm of Forms, for instance. Nor is it merely to say that we are necessarily corporeal. After all, most philosophers, and certainly not just phenomenologists, accept that that is the case. Instead, the phenomenologist is making the more radical point that our very notion of the world presupposes our corporeality. To have a world is necessarily to have a body, for no sense can be made of a world that could not be perceived from a certain location, with a particular set of sensory organs and so forth. There is, as Merleau-Ponty writes, 'a natal pact between our body and the world' (*PrP*: 6). In this sense, the body conditions everything I perceive and everything I could perceive, everything that could be touched, seen, tasted, heard or smelt. My body is, so to speak, woven into the fabric of that which is pre-reflectively given. It is, as Sartre puts it, 'co-extensive with the world, spread across all things' (*BN*: 318).

Clearly, the body is not being understood as an object here. On the contrary, Merleau-Ponty suggests that '[i]f objects may never show me more than one of their facets, this is because I am myself in a certain place from which I see them and which I cannot see' (*PP*: 92). The body as it is for me, the lived body, is therefore not an object, but 'our means of communication' with the world, 'that by which there are objects' (*PP*: 92). Hence the account we have been considering here differs from that of the materialist. To be in-the-world is to *inhere* in a world, but not as an object, whether material or otherwise. It is of course true that we can picture ourselves as objects, as human-shaped conglomerations of matter, say. But any such exercise presupposes a secret communion between world and body. So even as I try to picture myself as so much matter, I find myself already inhabiting a space which radiates outwards from me, surrounded by things that reveal themselves as at hand or as out of reach, or in other ways related to my fingers, feet, eyes and ears. My body, like anyone's, must be lived before it can be objectified.

Φ

To inhere in the world is, therefore, to find that the world reflects one's concerns as a corporeal being. The farmer's gate discloses itself as being a short walk away; the tree discloses itself to me as climbable; I see the knife's point as sharp enough to puncture flesh. And as one would expect, the corporeal nature of perception is particularly evident in our perception of other sentient bodies. Merleau-Ponty again:

> [I]t is precisely my body which perceives the body of another, and discovers in that body a mysterious prolonging of its own intentions, a familiar way of dealing with the world. Henceforth, as the parts of my body together compromise [*sic*] a system, so my body and the other's are one whole, two sides of one and the same phenomenon, and the anonymous existence of which my body is the ever-renewed trace henceforth inhabits both bodies simultaneously.
>
> (*PP*: 353–4)

I empathise with the child that has slipped and fallen because I know what it is like to feel the shock and humiliation of a sudden fall. When his mother rushes to comfort him, I know what her hug feels like too because I am a being that can hug and be hugged. My experience is partly constituted by my sense of what it is like to have a body; indeed, more generally, the corporeal character of perception is nowhere more evident than in our relations with others.

Merleau-Ponty concludes that all this 'makes another living being, but not yet another man' (*PP*: 354). But our concern is with the natural environment, and for that purpose we can afford to stop short at living beings and not concern ourselves with what it is to perceive a being as *human*. And the first thing to note here is that Merleau-Ponty seems to be correct: all that has been said so far does not just apply to our encounters with human beings. Observing a cat stalk its prey one feels the tension in one's own muscles, the effort to maintain balance and control. The thrill of watching the amazing leaps and turns of springbok in their escape from a predator is a visceral thrill. It is to feel the terror of an approaching lion and the wonderful freedom of having a body capable of leaping ten metres in a single bound. A pigeon takes off from the garden wall and then, its wings pulled back, plummets towards the ground. But then, in a moment, it has opened its wings and swept upwards into a nearby tree, its flight tracing an inverse parabola. I perceive the heart-stopping plunge to earth followed by the exhilarating recovery because, although I cannot fly, I nonetheless know the thrill of harnessing my momentum in air or water (cf. Abram 1996: 61).

Φ

To say that we are involved with a world in which we inhere is to say that, whether we are consciously aware of the fact or not, we are always already inhabiting the world. This applies generally, regardless of whether we find ourselves inhabiting worlds that have been shaped by human hands or ones that are, by contrast, predominately natural. Nonetheless, taking seriously our involvement with and inherence in the world prescribes a certain approach to environmental philosophy. In positive terms, it is to demand that in investigating our lived relations with the natural world we begin by interrogating our being-in-the-world. Expressed negatively, it is to reject any suggestion that we might be able to comprehend those relations by supposing, against Heidegger, Merleau-Ponty and the rest, that one's primary relation to the world is that of a disembodied subject confronted by a realm of objects.

To be sure, rejecting this supposition does not have any clear normative implications. One can reject the notion that one is at root a disembodied subject and yet, with complete consistency, not give a damn about conserving the natural world. Conversely, one can accept that notion and yet be moved to treat nature well. However, the fact remains that if the case set out in this chapter is correct, and if, moreover, we are to understand our lived relations with the natural world, be they harmonious or antagonistic or whatever, then we must begin by rejecting the notion that we are at root subjects confronted by a realm of objects.

But which approaches to environmental philosophy does this rule out? The surprising answer is, a great many of them. For much environmental philosophy is predicated on the notion that our most fundamental relations with the natural world are causal relations of the sort that have been, or could in principle be, described by science. Now this presupposition is entirely appropriate in the context of scientific inquiry. But when this conception of human-nature relations is vaunted as metaphysics, as some kind of scientific (or rather, scientistic) naturalism, we lose sight of what it means to inhabit the world, and so we lose sight of what it means to inhabit the natural world. On the one hand, we overlook our involvement with the world. For to regard the world as a realm of bare objects intrinsically bereft of meaning and value is to fail to comprehend the peculiar intimacy of the relation between the world we inhabit and our concerns. In more technical language, it is to suppose, wrongly, that our relations with that world are causal rather than intentional. On the other hand, to endorse scientistic naturalism is to lose sight of our inherence in the world. For although exponents of this kind of naturalism would no doubt reject any suggestion that we are essentially disembodied, their metaphysics commits them to the notion that the body is fundamentally an object present-at-hand. So although scientistic naturalism may be opposed to the radical mind-body dualism defended by some Platonists, for instance, it cannot do justice to the way in which the body is *lived*, the body which is internally related to our perception of the world and which therefore is, in Sartre's evocative phrase, 'spread across all things' (*BN*: 318). And in losing sight of *this* body, the proponent of naturalism finds herself unable to comprehend our inherence in the world.

In these ways, then, the proponent of naturalism overlooks our involvement with and inherence in the world. And so she finds herself unable to comprehend what it means to inhabit the natural world. In environmental philosophy and elsewhere, scientistic naturalism is a dead end. In the following chapters, we will take another route.

Notes

1. While agreeing with this general point, Erazim Kohák argues that in modern technological societies our lived experience has become so impoverished that the nature we experience has come increasingly to resemble our materialistic worldview (1984: 12–13). I discuss Kohák's views in Chapter 4.

2. Note 1 to Proposition 37 of Part 4 of his *Ethics*. On the environmental implications of Spinoza's philosophy, see Chapters 11–13 of Witoszek and Brennan 1999.
3. My use of the term 'involvement' differs from Heidegger's use of the term *Bewandtnis*, which is often translated as 'involvement' (see *BT*: 115). I am using involvement as a synonym of Heidegger's *Befindlichkeit*, which has been variously translated as 'attunement', 'situatedness' and 'state-of-mind'. On the difficulties of translating this term, see Polt 1999: 65.
4. It is interesting to consider whether this kind of questioning – questioning of something so fundamental to one's existence – is like other, more familiar kinds of questioning. To question whether a politician is telling the truth is one thing, but to question whether one is in-the-world is, one suspects, quite another – in part, no doubt, because this sort of questioning seems to presuppose that one is in-the-world in the first place. I consider some phenomenological responses to scepticism in Chapter 2.
5. One common assumption in these discussions is that it would necessarily be a bad thing to be alienated from nature. But this claim is open to question. After all, some writers have suggested that it is broadly speaking a good thing to regard the natural world as independent of and alien to human concerns (see further, Williams 1992). I discuss the various senses in which nature exists independently of human concerns in Chapter 5.
6. Cf. Heidegger: 'If, for instance, the south wind "is accepted" ["gilt"] by the farmer as a sign of rain, then this "acceptance" ["Geltung"] – or the "value" with which the entity is "invested" – is not a sort of bonus over and above what is already present-at-hand in itself – *viz*, the flow of air in a definite geographical direction' (*BT*: 111).
7. Interestingly, while Richard Polt admits that 'animals are tied to their "home" more tightly than we can ever be', he suggests that because human beings are capable of understanding themselves historically, in relation to a future and a past, they are 'capable of dwelling more intensely than an animal can' (1999: 78). (We will return to the topic of animal life in Chapter 2.)
8. In the following, I draw upon Merleau-Ponty's discussion of the 'body-subject'. But it is worth noting that much of that analysis was prefigured in Husserl's lectures on *Thing and Space*, and subsequently in *Ideas II* and *The Crisis of the European Sciences and Transcendental Phenomenology*. On the relation between Husserl's account of the body and Merleau-Ponty's, see Cerbone 2006, Chapter 4.

2
Animal Minds

2.1 Cross-species intersubjectivity

In Chapter 1 we saw that references to being 'in' nature can be interpreted in several different ways. Most of us know what it is like to be *au fait* with a particular natural environment, 'at home' in it. And regardless of whether we know what it is like to be in nature in this way, our perception will be conditioned by our inherence in the world. So whether or not we are at home in any natural environments, our perception of all phenomena, natural ones included, will necessarily be embodied – conditioned, that is, by the fact that we are of the same 'stuff' as the things we encounter.

Yet we can be in nature in other ways as well. In this chapter I examine the proposal that we inhabit an intersubjective world, containing at least some non-human subjects, and that we are therefore in nature in something like a social sense. I begin by arguing that there *is* such a thing as cross-species intersubjectivity, before moving on to show why this means that at least some non-human animals (hereafter 'animals') must be conscious. I turn next to the more practical question of how to interpret the methods and results of empirical studies of animal behaviour, and, in particular, the question of whether such studies can ever suffice to show that certain animals (or kinds of animal) are conscious. I argue that it would be a mistake to presume that such studies will only be able to yield one of the two results: either (a) that the animal in question is conscious in much the same way that we are conscious, or (b) that it is entirely non-conscious, like a stone. Drawing upon the work of Heidegger and Merleau-Ponty, I contend that this is a false dilemma.

Φ

Our topic, then, is cross-species intersubjectivity. But before considering this particular kind of intersubjectivity, I ought to say something about intersubjectivity in general. And to do this, as good a place as any to begin is with Heidegger's analysis of 'being-with' (*Mitsein*) (*BT*: §§ 25–7).

As we saw in Chapter 1, Heidegger points out that to exist as we do is necessarily to inhabit a world, to be 'in-the-world'. Among other things, this means that to describe the manner of being enjoyed by you or I, one must refer to certain items in the world, an 'environmental context of equipment' (*BT*: 154). So to describe how things are for me, I find that I have to refer to the pen I am holding, the paper underneath the nib of my pen, the book I hope to complete and so on. To describe these things and their various relations to me, moreover, I must refer to others, who are in certain important respects like me. The pen is *mine*, which among other things is to say that it might once have been someone else's property. It has a certain function: it is something that one writes with, where this 'one' refers to the anonymous collective denoted by the word 'anyone'. When I was very young, I learnt how to use it – presumably by imitating others. What is more, the pen was made by others – designed and manufactured by the people at Pentel. I'm using it to write a book which will be read by others – beings who, like me, are in-the-world. Indeed, writing is a recognised form of social activity: it is something one does, where, again, this 'one' refers to anyone, the collective Heidegger refers to as *das Man*. This should not be taken to imply that I first perceive certain items of equipment and then infer the existence of others. On the contrary, the actual or potential presence of others is integral to the presentation of these things. As Heidegger puts it, 'The Others who are thus "encountered" in [an] ... environmental context of equipment, are not somehow added on in thought' (*BT*: 154).

For Heidegger, then, our way of being is a being-in-the-world, which is necessarily a being-with others (*BT*: 156). This is not to say that we are essentially sociable creatures, since being-with is the precondition, not only for conviviality, but also for hostility and sociopathy. Nor does being-with require the actual presence of others. We are of course sometimes alone. Yet such moments count as moments of *solitude* or *isolation* on account of the actual absence of others who might have been present (*BT*: 156–7). Even Descartes' solitary meditations before his stove indicate a particular mode of being-in-the-world and hence a particular mode of being-with-others.

Φ

By means of analyses such as this, Heidegger aims to shed light on what he calls Dasein, our way of being. It is tempting to say that while his ultimate aim is to elucidate the enigmatic 'question of the meaning of Being', his first concern is with *human* being, and in support of this suggestion one could note that human beings do indeed typically exist as Dasein.[1] But the concept of Dasein is not coextensive with that of human being (Schatzki 1992: 82; cf. Thomson 2004: 401), and being-in-the-world is, to be precise, a being-with-other-Dasein, not a being-with-other-humans. Yet for all this, Heidegger fails to follow up the implications of his own reasoning, supposing instead that the token others one is 'with' (i.e., 'with' in an existentially fundamental sense) are all of a certain (human) type. For Heidegger, a being-with animals is not possible.[2]

Why exactly is Heidegger unwilling to admit the possibility of an interspecies being-with? Is such a relation unthinkable in the context of his account of *Mitsein*? Or could that account be extended to encompass our relations with non-human others?

One might think that the problem here is simply with Heidegger's choice of example. *Being and Time* does not abound with phenomenological descriptions of our relations with animals, but that, one might suppose, is simply because Heidegger chose to illustrate the fundamental structures of being-in-the-world by considering our manipulation of certain artefacts (the famous hammer of §15, for example). I also found this a natural place to begin, choosing to introduce the concept of *Mitsein* by describing my relations to my pen, my notebook and so forth. Here, as in Heidegger's own examples, the things are all artefacts and the others referred to all human. But what if I had chosen a different situation, one centred not on my manipulation of artefacts, but on my encounters with non-human beings?

Imagine, then, that I am contemplating these matters, not sat at my desk, but while out walking my dog.[3] To describe how I am is at once to describe my situation, and my situation is that I am walking through fields near my home, the sun on my back, relishing the space and freedom. Others are integral to my perception of the world. I am walking on the edges of the *farmer's* field. This is the sort of thing *one* does. As Heidegger pointed out, my walking must be understood against a background of shared, public understanding (cf. *BT*: 153). Yet this 'public' is not entirely human. My dog, Lucy, is also present, excitedly rooting around in the hedgerow – like me, relishing the sun. In fact, I perceive the sun as something in which we, Lucy and I, take pleasure. Lying in some nettles is a stick, bearing the tooth-marks that identify it as Lucy's. One could say that her presence is written into it. Indeed this entire stretch of field is part of Lucy's walk; her presence (or potential presence) is integral to my perceiving the field in the way I do. In this situation, the things with which I interact refer, not merely to other humans, but to at least one hairy non-human being as well.

It is tempting, on the basis of this example, to conclude that we should simply enlarge the boundaries of Heidegger's concept of *das Man*, to grant that the things with which we interact refer, not just to other human Dasein, but to a wider, interspecies public. But this conclusion should be resisted. The concept of 'the public', like Heidegger's concept of *das Man*, refers to a distinctively human domain, such that talk of an 'interspecies public' is oxymoronic. Even talk of 'communities', which in the wake of Aldo Leopold's work might seem more amenable to a cross-species interpretation, is inappropriate. No, my interaction with Lucy was precisely that: an interaction between one being and another, what we would usually call a face-to-face encounter. Efforts to conceive it in terms of the intimation of some kind of interspecies 'public' or 'community' are, at the very least, strained.

So how, as a phenomenologist, is one to understand my interactions with Lucy? Heidegger's analysis of *Mitsein* might not be the best place to start. For although in his accounts of solicitude (*Fürsorge*) Heidegger gestures towards an analysis of interpersonal, 'face-to-face' relations, his main concern is with impersonal social relations, or, more precisely, with the anonymous collective of *das Man*. As Sartre explains, 'Heidegger's being-with is not the clear and distinct position

of an individual confronting another individual ... but the mute existence in common of one member of the crew with his fellows, that
existence which the rhythm of the oars or the regular movements
of the coxswain will render sensible to the rowers and which will
be made manifest to them by the common goal to be attained, the
boat or the yacht to be overtaken, and the entire world (spectators,
performance, etc.) which is profiled on the horizon' (*BN*: 246–7).[4]

How, then, are we to understand my interactions with Lucy? It
would be misleading to say that the encounter involved a meeting of
minds. On the one hand, such talk evokes the radically individualis-
tic, Cartesian conception of human being that was (arguably) under-
mined by Heidegger's analysis. On the other, talk of a meeting of
minds is, to speak loosely, too 'mentalistic' to capture my interactions
with Lucy. My being-with Lucy, if it may be so described, involved
an intertwining of bodily intentions, a shared response of two lived
bodies to a common situation. It certainly cannot be understood on
the basis of a model that is merely cognitive.

To capture the bodily dimension of our relations with animals,
it may be helpful to leave Heidegger's analysis of *Mitsein* and turn,
once again, to Merleau-Ponty's account of the 'lived body'. To be
sure, Merleau-Ponty has little to say about our *social* relations with
animals, preferring the evidence generated from intrusive and alien-
ating laboratory-based studies (Behnke 1999: 99–100). However, his
account of interpersonal relations can, I think, be extended to deal
with cross-species interactions. The first thing to note in this regard
is that Merleau-Ponty explicitly pitches his account against those
who appeal to the image of one mind meeting another. 'To begin
with', he writes, '[others] are not there as minds ... but such for
example as we face them in anger or love – faces, gestures, spoken
words to which our own respond without thoughts intervening'
(*S*: 181).[5] It is true that Merleau-Ponty is referring to other humans
rather than to animals, but his analysis could be applied to animals
as well. So, for instance, with regard to my relations with Lucy, it is
true that she and I share an understanding, but that sharing takes the
form of an unspoken dialogue of expression and gesture, more like a
dance than an exchange of e-mails. Lucy's movements make sense to
me – I can, so to speak, read what they are saying, what she is saying;
and I respond to her in kind. She executes a 'play bow' (drops onto
her forepaws, tail up), and I respond to her unconsciously, bending

towards her, patting my knees. I mimic her movements, and she mine, through a kind of pre-reflective 'postural impregnation' (cf. *PrP*: 145; see also Abram 1996: 21, Caputo 1993: 126–7). For sure, I sometimes misread her behaviour, as she occasionally misreads mine, but the very possibility of such misinterpretations presupposes a basic mutual understanding between us, albeit one that can sometimes break down. And this basic understanding exists because, for all our differences, there is what Merleau-Ponty calls a 'kinship' between my body and Lucy's (*EP*: 190). We are corporeal beings inhabiting what is in some respects a common world.

If talk of a meeting of minds is inappropriate, then so too are appeals to inference. I do not perceive a Lucy-shaped body, only subsequently inferring that that body houses a mind, and I am sure that Lucy does not perform a similar inference in perceiving me (cf. Scheler 1954: 238–41). Nor indeed do I *assume* that the dog-shaped body I perceive houses a mind (against Kennedy 1992: 32). To speak of Lucy's mind is rather an imperfect way of expressing the fact that her bodily behaviour has a certain kind of meaning. Her mind is not some sort of ghost in the machine, the presence of which I infer from observing her movements. Her mindedness is manifested in her movements. Her behaviour does not provide *evidence* that she is conscious. I perceive her behaviour *as* conscious behaviour (cf. Gaita 2003: 52).

There is therefore no good reason to suppose that the others I am 'with' – that is, 'with' in an existentially fundamental sense – do not include at least some non-human beings. Lucy is not, so to speak, *over there* on the far side of an epistemological gulf, to be bridged by an argument from analogy or an inference to the best explanation (of which, more below). No, she is already *here*, in the Da of my Dasein (see further, Churchill 2008: 180–1).

2.2 The problem of animal minds

I am not the first to have suggested something along these lines. Following a remark from Derrida (1989: 57), Simon Glendinning speculates on an 'original Mitsein' between humans and animals (1998: 72). Likewise, John D. Caputo chastises Heidegger for having ignored the possibility of 'a kind of Mit-sein ... a way to be "with" animals' (1993: 127), while Charles S. Brown defends the notion of an interspecies 'being-with-others' (2007: 94).

All of these writers suggest that we are 'with' animals, in an existential sense, even if (as they all acknowledge) we are not 'with' them in the same way that we are typically 'with' our fellow humans, and even if they tend to assume – wrongly, as I have argued – that such interspecies interactions are best conceived on the model of Heidegger's account of interpersonal relations, rather than that of Merleau-Ponty.

Now if all this is correct, if a reworked conception of *Mitsein* can incorporate our relations with some animals, then we are presented with a third way to understand our place in nature. We can be in nature in the sense that we can be *involved* with it in certain ways. We are in nature in the sense that we *inhere* in it as corporeal beings. But we are also in nature in the sense that we find ourselves inhabiting an *intersubjective* world containing at least some non-human subjects.

Let us assume, for the sake of argument, that the last of these claims holds true, and that to describe the sort of being I am is to describe my relations not only with other humans, but also with certain non-human animals. If *Mitsein* incorporates some animals, must the animals that are incorporated possess phenomenal consciousness? Must there be *something it is like* to be them? Must they have *minds*?[6]

Some will contend that this does not follow. To be sure, the phenomenologist has told us something about how we perceive animals. She has shown that, being conscious ourselves, we tend to perceive animals' behaviour, or at least the behaviour of some animals, as conscious. But confined to the realm of human experience, a phenomenological investigation is unable to tell us anything about how non-human beings *themselves* experience the world. To do that, one would need a phenomenology of non-human perception, and for that one would need non-human beings capable of phenomenological reflection. But in the absence of any furred or feathered phenomenologists, one is not justified in inferring anything at all about what it is like to be an animal. In fact one cannot presume that there is anything it is like to be an animal at all. One cannot presume that any animals are conscious.

Φ

I should make it clear, right away, that I believe these objections are groundless. I will set out my reasons for this conclusion presently.

For now, however, it will suffice to note that the set of objections canvassed above implies a familiar sceptical problem. I call it the problem of animal minds.

This problem (if problem it is) presupposes the following account of our epistemic situation. I have immediate epistemological access to my own mind; I know, to speak loosely, that there is something going on in my own head. But how can I be sure that beings other than myself have minds? How, to be more precise, can I be certain that any of them have phenomenal consciousness? It must be at least logically possible that my dog, for instance, is nothing more than a complicated machine. The problem, then, is one of getting from *here* (my immediate awareness of my own mind) to *there* (the mind of another). Some sort of epistemological 'bridge' is needed, some way of justifying our belief in the consciousness of animals.

The argument from analogy has proved to be the most popular form of epistemological bridge.[7] So consider, by way of example, the claim of one professor of animal behaviour that although 'the private mental experiences [of animals], if they have them, remain inaccessible to direct observation', we can justifiably 'conclude that animals do experience suffering in ways similar to ourselves', so long as our reasoning is based 'on an analogy with our own feelings' (Dawkins 1980: 102). The reasoning here runs as follows: I know from my own case that a certain kind of observable event (e.g., my being pricked by a pin) is typically conjoined with a certain kind of mental state (a sensation of pain). Furthermore, I know that this mental state is typically conjoined with a certain kind of physical behaviour on my part (crying out). Now suppose that I observe a pin being pressed into a dog's flesh. I hear the animal's yelps and infer that the pin-prick and the yelp are mediated by a particular kind of mental state, a sensation of pain, just as they are in my own case. It is just that in this situation the pain 'happens' in the dog's consciousness, rather than in mine.

An argument from analogy is not the only way to try to prove the existence of animal minds. One can also employ an inference to the best explanation. According to this line of reasoning, the conclusion that animals have minds is thought to provide the best explanation of their behaviour, winning out over accounts that appeal to non-conscious processes only. As Marc Bekoff explains, the existence of a mind can serve as a 'hypothetical construct' and so provide the basis for a cogent explanation of the data provided by studies of animal behaviour:

> the 'privacy' of mental states does not necessarily present more or less of a problem for [the study of animal behaviour] than the invisibility of electrons does for chemistry. ... Claims about minds and electrons are posited as hypothetical constructs because they make sense of the data that has been collected.
>
> (Bekoff 1996: 234)

In my view, neither an argument from analogy nor an inference to the best explanation provides an adequate response to someone who doubts the existence of animal minds. I will explain why I think this below. For now, it is enough to note that framed in these terms the problem of animal minds is evidently a special case of a more general sceptical doubt – the problem of other minds. If I can only be sure that I am conscious, then it isn't just the consciousness of non-human others that is thrown into question. Perhaps the various human-shaped beings that surround me are also mindless mechanisms – 'zombies' or automata. Yet it would be a mistake to think of the problem of animal minds as nothing more than a special case of the more general problem of other minds. For there are reasons to think that the former is distinctive, that the attempt to bridge the epistemological gap between a human mind and the mind of an animal presents a special challenge. For one thing, the behaviour of animals, unlike the often more complex behaviour of humans, might appear to be explicable, at least in principle, in terms of non-conscious processes.[8] Scientists who adhere to 'Lloyd Morgan's canon' – the injunction against interpreting an action as 'the outcome of the exercise of a high psychical faculty, if it can be interpreted as the outcome of the exercise of one which stands lower in the psychological scale' – might therefore be reluctant to attribute consciousness to animals (Morgan 1894: 53). Another reason for doubting the existence of animal minds is that, unlike most humans, animals cannot tell us what they are thinking or feeling.[9] Consequently, evidence that they are thinking or feeling anything at all may be thought to be lacking (see, for instance, Banks 1995: 270; for further discussion, see Jamieson 1998: 91).

In the face of such difficulties, the sceptic concludes that there is always a chance that any particular animal is mindless and so merely a complicated machine or something of that kind. So whereas it might seem, to most of us, that at least some animals are minded, the sceptic maintains that this is open to question. This variety of scepticism – Donald Griffin refers to it as 'species solipsism' (1984: 28) – is often associated with Descartes, although its most committed advocate was probably Nicolas Malebranche, for whom animals 'eat without pleasure, cry without pain, grow without knowing it ... desire nothing, fear nothing, know nothing' (quoted in Harrison 1992: 219).[10] And it persists in the works of modern writers (see the views reported in Jamieson 1998: 100–1 and Rollin 2007: 270–2). Thus, to take just one example, the contemporary animal behaviourist J. S. Kennedy proclaims that 'although we cannot be certain that no animals are conscious, we can say that it is most unlikely that any of them are' (1992: 31).

Φ

So this is the problem to which we must respond. We need to determine whether, if we are indeed 'with' animals in an existential sense, this suffices to refute scepticism regarding animal consciousness. We need to investigate whether the phenomenologist's case for a cross-species *Mitsein* dispels the problem of animal minds.

To make any headway in this inquiry, we will need to consider what exactly it might mean to entertain scepticism regarding animal minds. And in doing this we will need to distinguish between some different kinds of 'species solipsism'.

The first kind is, one might say, purely intellectual. It may be logically possible that global solipsism is true, but it is certainly logically possible (and perhaps metaphysically possible too) that the animal-shaped beings with which one seems to share the world are all bereft of phenomenal consciousness.[11] This form of scepticism cannot be refuted, at least not by a phenomenological approach. The mere logical possibility that no animals are conscious cannot be dispelled. Yet one should not be too perturbed by this. To acknowledge that a particular strange state of affairs is logically possible is not necessarily to *doubt* what seems to be the case. I can acknowledge the logical possibility of my existing as a brain in a vat, yet I do not doubt that I am a living breathing person with two arms and two legs, a certain history, a distinctive circle of friends and family and so forth. The mere affirmation that a particular state of affairs is logically possible does not amount to sceptical doubt because, in short, doubt must be *lived*.

But what if one's scepticism is not merely intellectual but to some extent lived? What if one's scepticism involves real doubt? In order to answer this question it might help to imagine that we are dealing, not with scepticism regarding animal minds, but with global solipsism – scepticism regarding the existence of conscious others *per se*. And it might also help to imagine that we are actually debating these issues with a professed solipsist – I'll call her Jackie.

The problem with these imaginings, of course, is that the notion of debating anything with a genuine global solipsist seems absurd. For my part, I would want to ask Jackie why, if she genuinely believes that she is the only conscious entity in existence, she is trying to persuade *me* that solipsism is true. Who does she think she is trying to persuade? Given that she clearly believes she is trying to persuade someone other than herself, I would suspect that her scepticism is merely intellectual.

But what if Jackie actually doubts the existence of other minds? What if her scepticism is not merely intellectual? How then should we understand her claims? Clearly our story would have to be changed. If Jackie actually doubts the existence of other minds, then it is difficult to imagine her engaging in what we would call normal debating behaviour. If she is open to the possibility that I am conscious, she might respond to me, but only as if to an illusion. She would not be entirely sure that I really am a conscious being, and her behaviour would reflect this uncertainty. If, by contrast, she is absolutely convinced that no one is conscious other than herself, it is difficult, perhaps impossible, to imagine her engaging in any sort of debate at all. In fairy tales, boys and girls converse with teapots and cups, but only because these things have eyes and mouths and (it seems) mental lives. How could one converse with a being one believes to be non-conscious? And if Jackie not only doubts the existence of other minds, but the existence of the entire external world, her behaviour would be still harder to interpret. Perhaps Schopenhauer was right to claim that solipsism, when seriously espoused, can 'be found only in a madhouse' (1969: 104).[12]

Φ

So much, then, for the solipsist. Let us now turn to our second figure: the sceptic regarding animal minds. And as with our discussion of solipsism, let us begin with a concrete example.

So, imagine that while I am out walking Lucy, I bump into my (fictional) colleague, Sylvester. We get to discussing the topic of animal minds, and Sylvester tells me that he is unsure whether any animals have minds, whether, that is, there is ever anything it is like to be a non-human animal.

What are we to make of Sylvester's claims? One possibility is that they are the expression of a merely intellectual variety of scepticism. It is true that Sylvester will not be debating with Lucy; however, it might become evident that in speaking with me he is unreflectively comporting himself towards her as if to a conscious being. So, despite his professions of scepticism, Sylvester might stand and move in a way that indicates that he is unreflectively aware that the being snuffling around at his feet is conscious. Despite his vocal defence of 'species solipsism', Sylvester might seem to be 'with' her in an existential sense, and so his scepticism might seem to be merely intellectual rather than lived.

Suppose, however, that this is not the case. Suppose that as I am listening to his defence of scepticism regarding animal minds I become aware, through observing his body language, that Sylvester lives his scepticism, that he actually doubts whether Lucy has a mind. It dawns on me that Sylvester is not only espousing scepticism; his face, his arms, the way he is holding his body thus and so shows that, for him, Lucy might very well be a large hairy clockwork toy or something of that order.

It is difficult, but not impossible, to imagine someone relating to dogs in this way. Nicolas Malebranche, for instance, not only professed scepticism regarding animal minds, but also indulged a passion for vivisection, apparently without moral qualms. So perhaps he would have actually doubted that Lucy was conscious.[13] What cannot be doubted, I think, is that we are here dealing with a very odd form of life. Like lived global solipsism, lived species solipsism sounds more like an illness than a position.

I am not the first to have made such a diagnosis. Dale Jamieson, for instance, moots the possibility of a strange kind of autism that 'makes opaque the inner lives of animals, while leaving those of humans open to view' (1998: 95).[14] Of course, he adds, to maintain that the sceptic is autistic is no argument against scepticism, but merely 'name calling'. He is surely right about this. It would be mere name-calling to challenge scepticism by suggesting that the sceptic is suffering from some sort of mental dysfunction. But that is not what I am trying to do here. In linking lived scepticism regarding animal minds with mental dysfunction, my aim is rather to illuminate the existentially fundamental character of our relations with at least some animals. As we have seen, it would be quite wrong to say that I *assume* or *guess* that Lucy is a conscious being (*pace* Guthrie 1997: 55–6). My relating to animals like Lucy as if to conscious beings is more integral to who I am than talk of assumptions or guesses would suggest. And the same holds true of references to inference or intuition. Indeed one could say that when I contend, against Sylvester, that Lucy is a conscious being, I am giving voice to my existential kinship with beings like her, a kinship so fundamental to me that without it I would not be the kind of being I am. Or to put things in another way: it is impossible for me to be the kind of being that I am in the absence of bonds of the kind I share with dogs such as Lucy. Those bonds run deep – deeper than talk of assumptions and intuitions would suggest. Consequently, if someone doubts that Lucy is conscious, he is not doubting something I infer or assume to be true. Rather, his doubt indicates that his form of life is radically different from my own, so different that from my point of view it appears pathological.[15]

Φ

These arguments are not meant to prove that all those animals with which we have existential kinships are conscious or 'minded'. It remains at least logically possible that Lucy could turn out to be a sophisticated robot operated by remote control. However, if the case set out above is well taken, sceptical claims to the effect that *all* animals may be or are in fact mindless are either merely intellectual or else incomprehensible.

But even if these arguments are well taken, one might suspect that they leave all the important work to be done. For animals come in many different varieties, and it remains to be shown which *kinds* of animal are conscious. It remains to be shown to which kinds of animal an interspecies *Mitsein* might extend.

David Abram, for his part, adopts a very broad, animistic view, based on 'the intuition that every form one perceives – from the swallow swooping overhead to the fly on the blade of grass, and indeed the blade of grass itself – is an *experiencing* form, an entity with its own predilections and sensations' (1996: 9–10). He supports this bold claim by suggesting, in Merleau-Pontian mode, that 'one perceives a world at all only by projecting oneself into that world, that one makes contact with things and others only by actively participating in them, lending one's sensory imagination to things in order to discover how they alter and transform that imagination' (1996: 275–6, n. 3). This may be true. But it remains an open question whether, in any particular situation, one's projections are *appropriate* (cf. Clarke 2002). As we saw in Chapter 1, our perception of things is always shaped by our involvement with the world and our inherence within it. Even a blade of grass – to take Abram's example – discloses itself as it does because of the way it matters to us in the living of our lives and because of the various relations it bears to our bodies. Yet this is not to say that it is itself 'an experiencing form'. True, that claim may be endorsed in a merely intellectual way, as a hypothesis with a non-zero probability of being true. But to the extent that it is lived, it eludes my comprehension. Raimond Gaita argues that it is absurd to suppose that a stone has a 'rich inner life', not because there is overwhelming evidence that stones do not have thoughts, feelings etc., but because '[t]here is nothing which one can seriously suppose it would be like for [such] propositions to be true' (2003: 129). To think that a stone has a mental life is not to have failed to appreciate the significance of one or more pieces of evidence, but to have misapplied a concept. I would want to say something similar about Abram's claim that a blade of grass is an experiencing form. Derrida once confessed that he has felt embarrassed standing naked before his cat (2008: 113). That is understandable.[16] But if Derrida had instead admitted to feeling embarrassed before a blade of grass, his sanity would have been thrown into question. For my part, just as I cannot understand what it would be like to live one's life as if dogs were mindless mechanisms, so I cannot understand what it would be like to live one's life as if blades of grass were conscious. The two options strike me as being equally incomprehensible.

Φ

Once the animistic option has been rejected, one is left with the more interesting question of *which* animals are conscious. For, of course, it is often entirely reasonable to ask whether a particular animal (or kind of animal) is conscious. Sometimes this doubt may be a response to a lack of evidence – for example, one might doubt whether a certain species of animal is conscious because it lacks some anatomical characteristic, such as a nervous system. In these instances, one's assessment will depend, at least in part, on the results of physiological inquiries. In other cases, however, the key question will be how to interpret the animal's behaviour, and the appropriate methods of inquiry will be provided by sciences such as social ecology, comparative psychology and cognitive ethology.

Yet one might wonder how the results of science could possibly relate to the question of whether *Mitsein* can incorporate non-human beings. After all, grant, for the sake of argument, that certain people – most of us, no doubt – have a deep existential kinship with at least some animals. And grant also that this kinship, this 'cross-species *Mitsein*', cannot be understood in terms of our inferences, intuitions or assumptions. The fact remains that science does not deal in existential kinships of this sort. The aim of science is to investigate the world in an agent-neutral way. As soon as one appeals to a particular agent's privileged access to the world – to the lives of animals, perhaps – then, to be sure, one might be doing something interesting, but one will not be doing science. In the domain of religion there is a place for the testimonies of prophets and seers. In religious contexts, it is sometimes appropriate to place one's faith in the words of 'the chosen one'. But in science, things are different. In science, anything that can be demonstrated can, in principle, be demonstrated to all.

But this objection fails to convince. Doing empirical science well is at least in part a matter of learning not to misconstrue what one perceives. In this sense, science *does* rest on the testimony of particular agents – the testimony of scientists who have learned to perceive clearly, and whose observations can therefore be trusted.[17] And this holds true whether one is observing the behaviour of animals or the trails of particles in a bubble chamber. The good ethologist, like the good physicist, will have developed certain powers of observation. The good ethologist, for her part, will have learnt to read the behaviour of animals.

In order to see the connection between the scientific study of animal behaviour and the possibility of a cross-species *Mitsein*, we will need to consider what 'good observation' means in the study of animal behaviour. And the first thing to note here, I would suggest, is that when observing and interpreting *human* behaviour it is unwise to begin with the assumption that it is all non-conscious. So unless the ethologist – again, the good or expert ethologist – has firm evidence to the contrary, she will not proceed on the assumption that her animal subjects are mere mechanisms; indeed, if she were to do this she would be guilty, not of anthropomorphism, but of what Linnda R. Caporael (1986) dubs 'mechanomorphism'. Instead she will be open to the possibility that the animal in question is conscious and that at least some of the behaviour she is observing is conscious behaviour (cf. Spada 1997: 45). After all, as Jim Cheney and Anthony Weston point out, contrary to the common presupposition that an animal must 'prove itself' through its behaviour if it is to merit the ascription of some set of mental capacities:

> [T]he very necessity to 'prove oneself' may prove debilitating, may make its own satisfaction impossible. Precisely that demand already represents a way of closing ourselves off from the beings in question. Human beings trip over their own feet when treated in this way, and there is no reason to expect other animals to do better. ... Conversely, though it may seem paradoxical, removing the 'burden' of proof may be precisely what is necessary if it is to be met.
>
> (1999: 126–7; cf. Aaltola 2008: 189)

Indeed, they continue, 'unless we extend a basic courtesy to things [or animals] in our attempts to understand them, we cannot arrive at an understanding of them' (1999: 130).

Cheney and Weston's thesis accords nicely with some of the things working scientists say about the study of animal behaviour. So, to take one example, Barbara Smuts, an expert on baboon behaviour, maintains that 'treating members of other species as ... beings with potential far beyond our normal expectations, will bring out the best in them' (1999: 120). In fact, she continues, to understand anything about animals' mental lives, one must 'surrender' oneself to them, 'giving up control over them' (1999: 118; cf. Ruonakoski 2007: 77). For Smuts, as for Cheney and Weston, ethological studies will often require, not detached observation and methodological scepticism, but a certain kind of openness and empathy, a willingness to enter into the lives of animals.

Now, to be sure, this sort of approach will not always be appropriate (and we will consider some of the cases where it is not appropriate, below). But in many cases – when, for instance, one is studying the behaviour of mammals – it will be the best way to proceed. In these cases, the ethologist (again, the expert ethologist) will be open to the possibility that the animal is conscious; indeed, she may even adopt what Husserl called a personalistic stance towards it (see Mitchell 1997: 418–9).[18] This process is misunderstood if the human experimenter is conceived as an active agent who must solve the puzzle presented by the entirely passive animal subject. On the contrary, in many cases, the aim will be to regard the animal, not as a '[subject] of scientific inquiry' but as a 'fellow social [being]' (Smuts 1999: 113). The aim, in other words, will be to build up something like a socially mediated understanding of the animal. And this, I would suggest, is what 'good observation' in the field of ethology often involves – not quite *ob*servation (the prefix 'ob' signifies an alienation of observer from observed), but rather a willingness to approach the animal *as* a conscious subject, a kind of receptivity or openness, a particular kind of attention.

In trying to develop this understanding the experimenter begins, not with a *tabula rasa*, but with the interpretative framework that for her, as for all of us, has coalesced out of decades of lived, embodied social interaction. In the light of this framework, a particular movement might stand out as significant, similar perhaps to some human gesture. Further observation may confirm that the movement in question is significant (rather than accidental, say) and then, over time, it will be observed and interpreted again and again, set in the context of the animal's other behaviours, its physiology, environment and evolutionary heritage. After this process of interpretation, the movement may emerge as meaningful: not as *evidence* of consciousness, but as a *conscious movement* – as an angry swish of a tail, perhaps, or a contented purr.

If, through such investigations, the scientist comes to conclude that the animal *is* conscious, something quite interesting will have taken place. The animal's behaviour will now seem to make sense to her *as* behaviour, which (to switch idioms) is to say that claims to the effect that the animal is conscious will have taken on the status of hinge propositions for her. This is not to say that she will have become convinced of some truth that she formerly doubted. The transformation would have been more like a religious conversion: a transformation of the ethologist's basic form of life. In this way, the ethologist might even come to experience what Smuts calls 'a joyful intersubjectivity that transcends species boundaries' (1999: 114).

Once transformed, how will the ethologist regard claims that the animal in question is *not* conscious? Earlier in this section, I imagined myself responding to the sceptic's claim that Lucy is not conscious, and I concluded that if the sceptic were really to believe this, he would be suffering from some kind of mental illness. But the ethologist's reaction to the scepticism of her colleagues will hardly be this severe. She will judge her colleagues to be in error, of course. But at the same time she will, no doubt, appreciate that her coming to see the animal in question as conscious was the fruit of many hours of careful observation. And she will, no doubt, recall the fact that even she was at one time unconvinced that the creature had a mind.

2.3 The lives of animals

So far, we have been concerned with the question of whether animals are conscious – whether, that is, there is ever anything it is like to be an animal. We have not been investigating *how* things might be for an animal. Indeed it might seem that phenomenological inquiries would not be able to shed light on this second matter. After all, if we're doing phenomenology we are concerned with how animals are for us – how we experience them – not with how things are for animals. True, there could be some overlap between how things are for us and how things are for animals. So it could be that in revealing the former, phenomenological inquiries tell us something about the latter as well. But the fact is that phenomenological inquiries could never reveal whether any such overlap exists. Such inquiries might be able to reveal how we humans experience the world; however, animals' experience – whether it is like our own or unlike it – must remain mysterious, a big question mark.

Of course one might have similar concerns about our epistemic access to other *human* minds. Earlier in this chapter we considered sceptical worries about the consciousness of one's fellow humans (or one's fellow human-shaped beings, as things must seem to the solipsist). But a similar, if somewhat milder form of scepticism can be directed, not towards the claim *that* other minds exist, but towards claims to the effect that we can know *how* things are for these other minds. In other words, another sort of sceptic can concede that other minds exist, but reject any suggestion that we can (or could) know anything about how things seem to these other conscious beings (cf. Nagel 1974).[19]

We saw above how Heidegger's account of *Mitsein* can be brought to bear upon the first variety of scepticism. Yet it also militates against the second. For if Heidegger is right, then to be in-the-world, to exist as Dasein, is necessarily to find oneself, not only in a world of experiencing others, but in the midst of beings that, for the most part, share one's outlook on things. Thus in his 1929–30 lecture course, *The Fundamental Concepts of Metaphysics*, Heidegger maintains that human beings necessarily 'find themselves already transposed in their existence into other human beings, even if there are factically no other human beings in the vicinity' (*FCM*: 205). To be so transposed, he explains, is to be able to 'go along' (*Mitgang*) with what a being is and with how it is (*FCM*: 202ff.). So to say that each of us is always already transposed into the being of other humans is in effect to say that, in certain basic respects, we humans have a common outlook on the world.[20]

Is it possible to share such an outlook with an animal? True, we have already seen that we are 'with' some animals (or more cautiously, that there are no good reasons for concluding that this is never the case). But is it ever possible to 'go along' with them in their being? Do we ever find ourselves transposed into them?

Heidegger insists that this is not possible. To be sure, he concedes that animals are unlike stones in that '[w]e do not question that the animal as such carries around with it, as it were, a sphere offering the possibility of transposition' (*FCM*: 204). However, whereas we are always already transposed into the being of our fellow humans, we are never transposed into the being of 'the animal' (by which Heidegger means *any* animal).

The reason for this, Heidegger suggests, is that the world of the animal is qualitatively different from that of a human in that, while both humans and animals have a mode of access to beings, only the former are able to relate to beings *as* beings. So, for instance:

> When we say that the lizard is lying on the rock, we ought to cross out the word 'rock' in order to indicate that whatever the lizard is lying on is certainly given *in some way* for the lizard, and yet is not known to the lizard *as* a rock.
>
> (*FCM*: 198)

The animal has access to 'something' (and not even to 'something', corrects Heidegger, but rather to an unnameable '...'), '[b]ut this is something that *only we* are capable of experiencing and having manifest *as beings*' (*FCM*: 269).

To say that the animal has a mode of access to beings but is unable to apprehend them as beings is to say that it 'can only behave insofar as it is essentially captivated' (*FCM*: 239). To explain what this involves, Heidegger refers to an experimental study of the feeding behaviour of bees. Presented with a bowl of honey, a bee will suck for a period of time, before stopping and flying off. Yet, apparently, if the bee's abdomen is cut away while it is sucking, the animal will continue to suck indefinitely, the honey running out of it from behind. For Heidegger, this ghoulish result suggests that the bee's feeding behaviour is not triggered by its apprehension of the quantity of honey it is faced with: the bee does not recognise 'that it cannot cope with all the honey present' (*FCM*: 242). Instead, it stops feeding when its sucking behaviour is 'inhibited' by a state of satiation. This state 'cannot be registered in the bee if its abdomen is missing' (*FCM*: 242); hence the result. Heidegger's general (and extremely bold) conclusion is that 'the animal' (again, *any* animal) is surrounded by a 'disinhibiting ring', 'which prescribes what can affect or occasion its behaviour' (*FCM*: 255). Animals do not apprehend beings as beings and then respond to them; their behaviour is the result of certain patterns of 'disinhibition'. They are driven by their instinctual drives.

Φ

To say that we never find ourselves transposed into a being, in Heidegger's sense, is to say that we are unable, in principle, to appreciate how things are for it. It is true that to some extent and in a purely

intellectual way, we can understand how things are for the animal –
that is why Heidegger feels justified in speaking about the animal's
world. Yet if Heidegger is right, that world is not one that we can enter
into, imaginatively.

But this sweeping conclusion cannot be justified. It is true that
one can never know for sure how things are for an animal. As ever,
the spectre of merely intellectual scepticism cannot be exorcised.
However, the fact is that in the living of our lives we tend not only
to find ourselves in the presence of evidently conscious non-human
beings; in many situations we also find that we are already 'going
along' with these beings – in Heidegger's sense, already transposed
into them. Just as we do not need to infer *that* the animal is an experiencing being, so (in many cases) we do not need to infer *how* things
are for it. We share an understanding. To some extent, we share a
common outlook on things.

Of course this is not always the case. I share an understanding with
Lucy, which is to say, not only that there is no room for sceptical
doubts about whether she is conscious, but also that there is often
no room for doubts about how things are for her. (In fact, since she
is incapable of deception, there is less room for doubt here than
there would be if I were dealing with another human.) But in other
cases, and most of those with which professional ethologists deal,
the question of how things are for the animal may prove difficult
to answer. For, even if the ethologist comes to regard a particular
animal as conscious, there is no guarantee that she will be able to
read the *meaning* of its conscious behaviour. She might misread its
behaviour. Or she might find one or more of the behaviours of an
apparently conscious creature simply incomprehensible. The animal,
she concludes, is clearly *doing something*: it is clearly acting rather than
simply being acted upon by external forces. Yet its reasons for acting
remain mysterious.[21] It is interesting to note, however, that in these
cases the animal's conscious life is incomprehensible, not because the
animal's mind is a radically private realm, to which we are in principle denied epistemological access. The incomprehensibility is the
result of a hermeneutic failure, rather than an a priori epistemological
barrier. The animal defies our understanding because we are unable
to decipher the meaning of its manifestly conscious behaviour, not
because that behaviour is the visible sign of a mental cause that must
remain forever hidden from our view.

Φ

In some cases, then, the meaning of an evidently conscious behaviour will remain obscure. But in others it will not even be clear that one is observing *behaviour* rather than mere *movement*. One might try to regard the creature as conscious, but that interpretation might prove impossible to sustain. Indeed, one might be forced to conclude that the animal isn't conscious at all.

How are we to understand this result – the discovery that an animal is non-conscious? So far, we have been considering phenomenal consciousness, which means that we have been supposing that a being is conscious if and only if there is something it is like to be that being. But this Nagelian stipulation encourages one to think that an ethological inquiry, if it reveals anything at all that is relevant to the so-called 'problem of animal minds', will only be able to reveal one of two things: (1) that there is something it is like to be the animal in question, in much the same way that there is something it is like to be you or I; or (2) that there is nothing it is like to be the animal, just as there is nothing it is like to be a stone.

Are these the only options? As we have seen, Heidegger would have contended that they are not. Animals are unlike stones in that they have a mode of access to beings, but they are unlike humans in that they are unable to apprehend beings as beings.

Yet Heidegger's account is overly simplistic. Never one to do things by halves, he proclaims that he is inquiring into the 'essence' of animals and thus into something that 'holds true of all animals' (*FCM*: 186). But are we to believe that all animals, from starfish to star-nosed moles, share what is essentially the same mode of access to beings? Many writers have contended – rightly, in my view – that this is not the case at all. In particular, there is evidence to suggest that at least some animals *can* apprehend beings as beings. Consider, for instance, the phenomenon of tool-use in the great apes. When a chimpanzee sees a branch it may, in some circumstances, be apprehending the branch *as* a potential tool.[22]

In the man's defence, it may be noted that Heidegger's aim in his 1929–30 lecture course is not so much to elucidate the being of animals *per se*, as to shed light on our way of being (*FCM*: §63). More precisely, he wants to show that, tranquillised by our blind allegiance to the dictates of *das Man*, inauthentic individuals are acting more like bees than men (Kupperus 2007: 21). But these issues need not detain us here. For present purposes, it will suffice to note that Heidegger would have rejected the dilemma cited above. It is true that he would have rejected any reference to consciousness, yet his account suggests that it is *not* the case that animals are either conscious like us or else not conscious at all. Other options are available. An animal might have a mode of access to beings, but one that is qualitatively different from the access enjoyed by you or I.

Φ

Like Heidegger, Merleau-Ponty would reject the dilemma cited above, though for different reasons. He concedes that some animals resemble machines – 'animal machines', like the medusa (roughly, a jellyfish). True, there *seems* to be a unity to the behaviour of these creatures, a *Bauplan*, but he claims that this 'has only a descriptive meaning' (*N*: 169); it reflects nothing more than how we see the animal. For physiological investigations reveal that where we see unity there are only independent movements. We see an animal behaving in various ways, moving itself in such and such a direction, gathering food for itself, and so on. Yet in truth there is simply one part of the animal moving another part, one part gathering food for another part and so on. To be sure, these various movements are closely coordinated; however, this coordination is not effected by any unified agency. If the various movements of the animal fit together, this is on account of the evolutionary history of the creature and nothing more. As such, the animal 'resemble[s] the executor of a plan traced from the outside' (*N*: 176). There is no real unity to it, just a set of externally related mechanical systems.

Animals like the medusa resemble machines in that they have no mode of access to beings, they 'take account of nothing' (*N*: 169). By contrast, other animals have what Merleau-Ponty calls an 'operating meaning' (*N*: 163) or, following the early twentieth-century biologist Jacob von Uexküll, an *Umwelt*. This, Merleau-Ponty explains, is not 'a sort of quasi-interiority' (*N*: 157), a private sphere of consciousness or proto-consciousness. The concept of the *Umwelt* is rather an attempt to capture the dynamic interplay between the organism's behaviour and its world. It is 'the world implied by the movement of the animal, and that regulates the animal's movements by its own structure' (*N*: 175).[23]

Merleau-Ponty also endorses von Uexküll's view on the distinction between lower animals and higher animals. Lower animals rest in their *Umwelt* 'more surely than the infant in its cradle' (*N*: 170). The urchin, for instance, is so well adapted to its surroundings that they cease to be surroundings at all: 'it lives truly as if there were only a world and only an urchin' (*N*: 171). Higher animals, by contrast, are presented with a *Gegenwelt*, a 'world in counterpoise' (*N*: 170). They have a nervous system which 'amounts to a rejoinder to the exterior world ... a "copy", a 'mirror of the world' (*N*: 171). This means that the exterior world affects them, not as a cause, but as a sign (*N*: 171).

Such, then, is the case with higher animals. Humans, however, represent a different case again. '[T]he dog, and *above all* the human, moves', writes Merleau-Ponty (*N*: 209, my emphasis). This is not simply to say that the *Gegenwelt* of humans is richer and more complex than that of a dog, say. On the contrary, Merleau-Ponty insists that 'The relation of the animal to the human will not be a simple hierarchy founded on an addition [e.g., human beings conceived as rational animals]: there will already be another manner of being a body in human being' (*N*: 214). To be human is not merely to be presented with a *Gegenwelt*, but to have a conception of an objective world (*PP*: 327; see also *PP*: 78).

Φ

These are difficult ideas which deserve a more thorough treatment than I can give them here. What is more, I do not intend to appraise them. I do not intend, for instance, to assess whether the relevant analyses of animal behaviour have stood the test of time. The important point to note, for our purposes, is the way in which Merleau-Ponty's account bears upon our initial dilemma. We began by supposing that an ethological inquiry, if it reveals anything at all that is relevant to the so-called 'problem of animal minds', will be able to yield one of only two results: either (1) that the animal in question is conscious in essentially the same way that we humans are conscious, or (2) that the animal is entirely non-conscious, like a stone. We saw that Heidegger provided a third option: that the animal might be 'poor in world', able, that is, to access beings, but unable to apprehend them *as* beings.

Merleau-Ponty's account complicates matters further. Against Heidegger, he maintains that behaviour comes in four qualitatively different varieties. Some animals (e.g., medusae) are essentially machine-like. Others are unlike machines in that they have *Umwelts*. Still others possess a nervous system and hence 'a mirror of the world', a *Gegenwelt*. And humans, finally, represent a different case again. Our original dilemma has become a quad-lemma. Of course a number of questions remain. Are there just four categories here?[24] How could one determine which category a particular animal belongs to? But what is, I think, evident is that just as reality cannot be exhaustively partitioned into *res cogitans* on the one hand and *res extensa* on the other, so animals do not come in just two varieties, those that are conscious in much the same way that we are conscious, and those that are mindless mechanisms. The sceptic is right to think that the conscious lives of animals are often opaque. But she is wrong to think that they are either conscious like us or else mindless mechanisms. Matters are more complex and more interesting than that.

Notes

1. 'Typically' because not all humans exist as Dasein. Babies don't, for instance. Though Heidegger has little to say about childhood development, Merleau-Ponty, a one-time Chair of Psychology and Pedagogy at the Sorbonne, does. See especially his essay, 'The Child's Relations with Others' (*PrP*: 96–155).

2. Rather confusingly, Heidegger occasionally writes that we are 'with' animals in the sense of being-with, though not in the sense of 'existing-with' (*FCM*: 210). Nonetheless, his basic point remains the same: however we are 'with' our fellow humans, our being with animals is different in kind. I will say more about Heidegger's views on 'animal being' below.

3. On the phenomenology of our relations with dogs, see Shapiro 1999.

4. Interestingly, Heidegger came to accept Sartre's view that the analysis of *Mitsein*, as presented in *Being and Time*, does not do justice to the phenomenology of interpersonal relations (Zahavi 2001: 158, n. 7). For my part, I do not mean to suggest that the *only* reason Heidegger's account cannot do justice to my relations with Lucy is that it cannot acknowledge what Sartre calls my concrete relations with others. On the contrary, there are a number of reasons why Heidegger's analysis fails to account for our relations with animals (see further, Calarco 2008).

5. Merleau-Ponty is here describing Husserl's mature view, which on this topic he endorses.

6. For the moment, I take 'has a mind' to be synonymous with 'is phenomenally conscious'. So I am stipulating that if a being has a mind, then there is something it is like to be that being (even if we could never know what exactly it would be like) (cf. Nagel 1974). This stipulation is, however, merely provisional (merely 'for the moment'). Towards the end of this chapter, I discuss some of the shortcomings of this Nagelian formulation.

7. For a historical example, see Romanes 1882.

8. Though Skinnerian behaviourists would no doubt contend that even human behaviour could, at least in principle, be adequately explained in this way.

9. In view of recent studies of certain great apes, this, arguably, should read '*most* animals' (see, e.g., Savage-Rumbaugh et al. 2001; Fields et al. 2005).

10. The view that Descartes denied that animals have minds has been challenged by Cottingham (1978). Cottingham's defence is rebutted in Harrison 1992. Harrison's measured conclusion is that Descartes 'did not adamantly insist that animals could not feel ... but rather showed that there are no irresistible reasons for asserting that they do' (227).

11. Like Wittgenstein, Merleau-Ponty thought that the very notion of global solipsism, or at least the notion of *defending* global solipsism, was incoherent (witness his remark that 'I can evolve a solipsist philosophy, but in doing so, I assume the existence of a community of men endowed with speech, and I address myself to it' (*PP*: 360)). But 'species solipsism' presents a different case. It may be incoherent to doubt the existence of others *per se*, but it is not obviously incoherent to doubt that any non-human animals have minds.

12. For a phenomenological account that treats scepticism in general as a condition in need of treatment, rather than as a position in need of refutation, see George L. Kline's discussion of the work of Gustav Shpet (1996: 158–9).

13. Or perhaps not. It is interesting to note that although Malebranche used to beat his dog there is no record of him beating stones. But if dogs and stones are both non-conscious, why the difference in behaviour? (cf. *N*: 166.)

14. This would be a *very* strange form of autism. In fact, there is some evidence to suggest that at least some autists have a special ability to read the behaviour of animals. See further, Grandin and Johnson 2006.

15. In a Wittgensteinian idiom, one could say that the proposition 'Lucy is conscious' is, for me, a hinge proposition (cf. remarks 341, 343 and 655 of *On Certainty* (1969: 44, 87)). For a slightly different response to the problem of animal minds, see James 2009.

16. At least, I can understand the feeling. By contrast, Merleau-Ponty writes that 'A dog's gaze directed towards me causes me no embarrassment' (*PP*: 361).

17. Cf. David Edward Shaner's suggestion that part of what made Louis Agassiz a great biologist was the 'feel' he had for nature, and in particular his eye for the details of natural phenomena (1989: 177–8).

18. It is true that Husserl does not attribute personhood to non-human animals. However, as Corrine Painter has argued, there are good grounds for thinking that a personalistic stance, in a broadly Husserlian sense, can and should be adopted towards some animals (2007: 101).

19. At the limit, this second form of scepticism transforms into the first. For if one has absolutely no sense of how things are for another being, on what grounds could one conclude that that being is conscious?

20. Heidegger would no doubt reject 'subjectivist' talk of 'outlooks'; but for the sake of convenience, let us stick with this idiom for the moment.

21. To be sure, if all the creature's behaviour proved incomprehensible, one could not relate to it as if to a conscious being. Here, however, we are considering a case where one or more specific behaviours of an apparently conscious being are incomprehensible.

22. But compare Merleau-Ponty's commentary on a 'monkey' using a branch as a stick: 'having become a stick for the monkey, the tree branch is eliminated as such – which is equivalent to saying that it is never possessed as an instrument in the full sense of the word. ... For man, on the contrary, the tree branch which has become a stick will remain precisely a tree-branch-which-has-become-a-stick, the same *thing* in two different functions and visible for him under a plurality of aspects' (*SB*: 175). See further, Embree 2008.

23. To explain the concept of the *Umwelt*, Merleau-Ponty follows von Uexküll in employing an analogy with music. An *Umwelt*, he explains, is rather like 'a melody that is singing itself', for while, in one obvious sense, we sing the melody, it also 'sings in us', in the sense that the notes themselves demand a certain kind of resolution (cf. *N*: 173–4; *PP*: 78). Similarly, just as an animal creates its *Umwelt* through its behaviour, the *Umwelt* itself demands certain behaviours on the part of the animal.

24. Iain Thomson distinguishes 'a *continuum of Dasein*', stretching from 'worldless organic matter' (grade 1) to 'entities with even richer worlds than human *Dasein*' (grade 7). His point, he stresses, is not to provide an exhaustive account of the various kinds of Dasein, but 'simply to suggest that we could articulate "degrees of *Dasein*" with more subtlety than Heidegger himself ever did' (2004: 402–3).

3
Nature's Value and Other Obsessions

3.1 How nature matters

As we saw in Chapter 1, we are always already involved with a world. For anything to 'show up' for us in experience it must bear upon our lived concerns in some way; in a very general sense, it must *matter* to us. Heidegger expresses the point by saying that involvement (*Befindlichkeit*) is a basic existential structure, which is to say that we are always attuned to the world in such a way that things disclose themselves as mattering to us (*BT*: 176). This involvement, moreover, is always accompanied by a particular affective tone: in Heidegger's terms, whatever shows up for us in experience does so in the light of some mood (*Stimmung*) (*BT*: 173). So, for instance, to be in an elated mood is to inhabit a world that is fizzing with possibilities. To be depressed is, among other things, to find oneself in a world without hope. Indeed Heidegger suggests that even a relatively detached contemplation of objects present-at-hand is not so detached that it is without its characteristic mood, a 'tranquil tarrying alongside' (*BT*: 177).

Even at this early stage in our discussion, we can see the distance between the phenomenologist's account and certain metaethical positions. According to one such position, principally associated with Hume, the world is fundamentally a realm of things that do not matter to us – a value-free realm, as this is usually put. True, we experience things as mattering to us in certain ways; we see them as being attractive or repellent, valuable or worthless and so on. But in the final analysis these properties have their source not in the things themselves but in our subjective states. So although the mug of coffee has certain objective properties of shape and mass and so on, its attractiveness is not one of them. That property reflects nothing more than my desire for a sip of coffee; it has its ultimate source in me, not in the world.

But for phenomenologists it is entirely wrong to suppose that the world is gilded and stained 'with the colours borrowed from internal sentiment' (Hume 2004: 135). As Heidegger insists, having a mood is 'not itself an inner condition which then reaches forth in an enigmatical way and puts its mark on Things' (*BT*: 176). It is not just that things necessarily disclose themselves to us in the light of some mood, but that no sense could be made of a world in which things did not so disclose themselves. Moods make 'it possible first of all to direct oneself towards something' (*BT*: 176).[1] Moreover, the reason why no sense could be made of a world that is not encountered in the light of some mood is because, far from cutting us off from the world, moods actually *constitute* the world. One does not first apprehend a world and then respond to it by adopting a particular mood. Instead, whatever discloses itself as a constituent of the world does so because of the mood into which one has been, as it were, swept up. A mood, then, 'comes neither from "outside" nor from "inside", but arises out of Being-in-the-world' (*BT*: 176). In this respect, and as the following passage from David Abram nicely shows, talk of moods blurs the boundary between the world and the subject:

> My life and the world's life are deeply intertwined; when I wake up one morning to find that a week-long illness has subsided and that my strength has returned, the world, when I step outside, fairly sparkles with energy and activity: swallows are swooping by in vivid flight; waves of heat rise from the newly paved road smelling strongly of tar; the old red barn across the field juts into the sky at an intense angle. Likewise, when a haze descends upon the valley in which I dwell, it descends upon my awareness as well, muddling my thoughts, making my muscles yearn for sleep.
>
> (1996: 33)

Φ

All this applies to the world in general and so to the natural world as well. Nature discloses itself to us only in so far as it matters to us.

But this is not to say that nature necessarily discloses itself as *valuable*. For in order for a thing to count as valuable it must (or should), not just matter, but matter in some *positive* way to a valuer, which is typically to say that the valuer must (or should) care about the thing to some extent.

To bring out the difference between the general kind of mattering indicated by references to *Befindlichkeit* and valuing, consider the following passages:

> Graceful animals like gazelles and the antelopes spent their days in abject terror while lions and panthers lived out their lives in listless imbecility punctuated by explosive bursts of cruelty. They slaughtered weaker animals; dismembered and devoured the sick and the old, before falling back into brutish sleep where the only activity was that of the parasites feeding on them from within. Some of the parasites were hosts to smaller parasites, which in turn were the breeding ground for viruses. Snakes moved among the trees, their fangs bared, ready to strike at bird or mammal, only to be ripped apart by hawks. ... [A]s he watched, the unshakeable conviction grew that, taken as a whole, nature was not only savage, it was a repulsive cesspit. All in all, nature deserved to be wiped out in a holocaust – and man's mission on earth was probably to do just that.

> All jungles are evil. ... [T]he first rule of the jungle is fear, and then hunger and thirst. There is fear everywhere: in the silence and in the shrill calls and the wild cries, in the stir of the leaves and the grating of branches, in the gloom, in the startled, slinking, peering beasts.

The first quotation is from Michel Houellebecq's *Atomised* (it expresses the boyhood views of the character, Michel) (2001: 38). The second expresses the view of the narrator in Leonard Woolf's novel *The Village in the Jungle* (2006: 3–5). Much could be said about these passages, about the fact that they both express Europeans' views on exotic fauna, for instance. But for now it will suffice to note that the natural world *matters* to both of these individuals. They are both attuned to nature, in Heidegger's sense. What is more, both their encounters are shaped by certain moods – in the first instance, by repugnance; in the second, by fear. Yet neither boy nor man *values* nature.

A third quotation, this time from a horrified Luc Ferry:

> The photograph shows a bull being followed by a disturbing crowd. It is immediately apparent that he is going to die. But as the scene evokes a chilling ritual, we sense that the road to this final libera- tion will be long and painful. This strange dream unfolds in our time, in Spain, in a little village called Coria. It is a game, of course, the rules of which are extremely simple: at four in the morning, a bull is set loose on the streets, where he is riddled with darts, first in the eyes and most sensitive parts. Four hours later, the animal is beaten until he dies of his wounds. In the picture, he looks like a pin cushion. The little white dots are so close together he almost looks snow-covered. Two men are pointing at him, smiling. One of them holds a banderilla which he will plant, several seconds later (as appears in another photo), in the animal's anus.
>
> (Ferry 2008: 147)

As Ferry notes, the animal inspires such fascination precisely because it is *not* regarded as a mere thing (2008: 148). The ritual only makes sense because the animal *matters* to the villagers in something like the way other humans matter, which in phenomenological terms is to say that this cruelty is an expression – albeit a perverted one – of our being-with animals. But although the bull matters to the villagers, they nonetheless do not value it.

Φ

The very general kind of mattering indicated by references to 'involvement' or *Befindlichkeit* does not, therefore, amount to valuing. Whatever shows up for us as a constituent of the world must do so because it matters to us in a very general sense; however, not every- thing that discloses itself in this way will show up for us as valuable (see further, Thomson 2004: 389).

How does this bear upon morality (or 'ethics' – I will use the terms interchangeably)? Can something matter to us morally but in a way that cannot be understood in terms of the value it has or the manner in which it is valued?

It is typically assumed, and not just by consequentialists, that this is not possible, and that all moral considerations can be adequately expressed in terms of value. Thus, in his popular textbook on environmental ethics, Joseph R. Des Jardins writes that

> central to a comprehensive environmental philosophy is a consideration of the nature and scope of value. A full account of value *determines* the ethical domain
>
> (2001: 132, my emphasis)

Similarly, in his recent introduction to ecological ethics, Patrick Curry maintains that 'there can be no ethics without value' (2006: 40).

These claims might seem obviously true. But they are not true at all. Indeed, I will contend in what follows that at least some of our moral relations with the natural world cannot be adequately understood in terms of environmental *values*. I will argue, in other words, that in environmental ethics we should resist the urge always to think, speak and write in terms of value.

3.2 Technology

We will consider the topic of environmental values presently. But before this we need to consider the alleged shortcomings of 'values-thinking' in general. And in order to do this we will need to engage with the work of Heidegger, a man who believed that 'through the characterization of something as "a value" what is so valued is robbed of its worth' and (rather more hysterically) that 'thinking in values is the greatest blasphemy imaginable against Being' (*BW*: 251). In particular, we will need to examine his account of *technology*.

In considering Heidegger's views on this topic, the first thing to note is that he uses 'technology' (*die Technik*) as a term of art. For one thing, the 'technology' of which he writes is not a particular kind of artefact. To speak of technology, he stipulates, is not to refer to any kind of being, but to the way in which beings reveal themselves to us in the first place. So it is not to speak of one sort of being (human-made ones, say) as opposed to some other sort. Instead, it is to refer to the basic way in which all beings, artefactual and natural, reveal themselves to us as beings. In Heidegger's idiom, 'technology' signifies the 'way of revealing' (*BW*: 318) or the mode of 'Being' that holds sway in the modern era. And to say that

that era is dominated by technology is therefore to say that in the modern era things tend to reveal themselves technologically.

In Heidegger's view, anything that reveals itself technologically does so as 'standing reserve' (*Bestand*) (*BW*: 322), as, so to speak, 'standing by', ready to be put to use. In more familiar language, one could say that it reveals itself as a mere resource, as something to be used and nothing more. So when he contends that the modern age is dominated by technology, Heidegger means that nowadays things tend to reveal themselves to us as means to ends: they're there to give us a 'kick', for instance, or to make us look good, or to increase productivity, or to enable us to get from A to B. They show up in our lives only to the extent that they are useful.

All this is best illustrated by means of an example. So, consider a visit to a supermarket. The items on the shelves are all *for* something – to satisfy our hunger and thirst, to give us pleasurable experiences, to make us smell nice or look attractive, to reduce cholesterol, to clear up this or that kind of stain or spill and so forth. The same may be said of the people who serve us. Mrs X or Mr Y stack shelves and man the checkout, respectively; indeed from certain points of view they are *for* these roles: for the company, for instance, they are 'human resources' (cf. *BW*: 323). This particular supermarket, moreover, exists for the good of the company that owns it. For Tesco it is nothing more than a profit generator. Yet such big business is needed, we are told, for the benefit of The Economy, in order to maintain a sufficiently high GDP. And all things considered, a thriving economy is there for us – for our benefit and for that of future generations of obedient consumers. And as for us? Well, our function is to keep craving and spending, for consumer demand is one of the things that keeps the whole technological system in motion. Everything, in short, is there for something else. In Heidegger's terms, everything is *Bestand*.

Two points need to be emphasised here. First, Heidegger stresses that technology may be associated with certain norms. In a technological world, he maintains, the primary end is that of efficiency, and the primary task is that of managing the world so that things may run more smoothly. Accordingly, the most esteemed mode of thought is the 'calculative' variety needed to ensure the most efficient use of resources.

The second point that needs emphasising is that, for Heidegger, technology is on the rise, which is to say that things are *increasingly* coming to reveal themselves to us as standing reserve, as mere resources. Now the possibility that *everything* might come to reveal itself as standing reserve is, for Heidegger, certainly deplorable. However, there is more to his position than the fear that modern culture might come to have a one-dimensional and deficient view of the world. The 'supreme danger' to which Heidegger refers is, one might say, metaphysical, for he attributes to the technological mode of revealing the peculiar and dangerous power to encroach into all areas of life, driving out non-technological understandings as it does so (see *BW*: 331 *ff.*). The drive behind this expansion is the drive towards the relentless ordering of standing reserve in the pursuit of the goal of efficiency. But total efficiency is a nebulous goal (if it can be considered a goal at all), and indeed Heidegger maintains that the drive to order all things in the most efficient way is not like an *ambition*, a conscious drive towards some clear and distinct end point. On the contrary, it is, he says, a manifestation of sheer willing for willing's sake, the exercise of the power to order, simply 'because...' Heidegger's fear is that the technological will might eventually come to extinguish all other modes of revealing, that it might come to impose itself as 'the presumed unique mode of disclosure' (Haar 1987: 86), and that 'calculative thinking may someday come to be accepted and practiced *as the only* way of thinking' (*DT*: 56). In such a nightmarish future, all would have been sacrificed to the modern idols of efficiency and management, and the world would have become a featureless expanse of standing reserve, a domesticated world drained of all richness and depth.

Φ

To convey the pernicious effects of technology, Heidegger frequently refers to our exploitation of the natural world. Revealed technologically (or in Heidegger-speak, 'enframed'), dandelions become 'weeds', old-growth forest becomes 'timber', grey squirrels and badgers reveal themselves as 'pests' to be culled. 'The earth', he writes, 'now reveals itself as a coal mining district, the soil as a mineral deposit' (*BW*: 320). Nature has become 'a gigantic gasoline station, an energy source for modern technology and industry' (*DT*: 50). Nowadays, he laments, soil, rivers and woods are all *Bestand* – insofar as they matter, they do so only because they are useful.

To consume a resource is, by definition, to use it. But it is not necessarily to use it up. So if Heidegger sees technology as being manifest in
the drive to consume 'natural resources', he also thinks it is evident in
our tendency to regard the natural world as a source of what economists call non-consumptive use-values. For example, in 'The Question
Concerning Technology', Heidegger bemoans the fact that the Rhine
has come to disclose itself 'as an object on call for inspection by a
tour group ordered there by the vacation industry' (*BW*: 321). To be
sure, the river is not used up by the tourists (though Heidegger would
say that it is not allowed to disclose itself as the wonderful 'thing' it
in fact is). Yet it is nonetheless consumed by the tourists' eyes; it is
subject to what John Urry (1995) has called 'visual consumption'.
For Heidegger, the sublime subject of Hölderlin's hymn has become
nothing more than *Bestand*, a spectacle to be captured on film or in
JPEG files.

Were he alive today, Heidegger would no doubt observe that the
workings of technology are evident, not merely in efforts to exploit
the natural world, but also in the words and actions of those who
seek to conserve it. Consider, for instance, the following statement
from Sir Colin Marshall, onetime Chairman of British Airways:

> The tourism industry has, in reality, just one basic product – the
> environment of planet Earth. Our job is to deliver it, through trans-
> port, accommodation, resource management and other services, to
> customers who wish to 'hire' the environment of beaches, oceans,
> mountains, forests, lakes and cities for a short while. It is, therefore,
> essential that travel and tourism interests should invest heavily in
> environmental care and conservation, if we expect the industry to
> have a productive, long-term future.
>
> (quoted in Neale 1998: viii)

Marshall supports environmental causes, but he makes his case
using a singularly 'technological' vocabulary. And in this he is, of
course, not alone. Environmental problems are frequently thought
to involve the proper *management* of natural *resources*, and resolved
using the exemplary 'calculative' device of cost-benefit analysis.
Decisions are made on the basis of monetary valuations of ecosystem 'services' or, in other cases, 'functions' (e.g., Ansink et al. 2008).
Concerned members of the public are beaten into submission by

environmentalists reeling off statistics – rates of carbon dioxide emission, percentages of recycled waste by country and so on. We are told by the champions of green business that '[t]he environment is our core asset, the key component of product quality and an increasing priority for our consumers'[2]. This list could be extended; and indeed the desire to 'explain' a phenomenon by providing a bullet-pointed list of its characteristics exemplifies the 'calculative' mode of reasoning with which technology may be associated.

Φ

This is merely a sketch of Heidegger's account of technology. In particular, it does not present the *reasons* Heidegger provides in support of his views. I will say something about these reasons below. For the moment, however, I would like to draw attention to the relation between Heidegger's account of technology and his criticisms of 'values-thinking'.

The first thing to note in this connection is that for Heidegger technology blinds us to the various ways in which things can matter. So whereas in past ages (he claims) the world was infused with aesthetic, religious and moral significance, the modern world is (or is becoming) 'flat and devoid of background', a bland field of standing reserve (*QCT*: 142).[3] In a technological world, things matter, but in only one way; they only matter to the extent that they are useful. The possibility that they might disclose themselves in other ways – as things of beauty, for example – has been eradicated, driven out. One could say that when things are forced to reveal themselves in this constricted way they are *violated*, that *violence* is done to them. In his excellent book on Heidegger's later thought, Julian Young writes that 'Violation is, in one way or another and to one degree or another, preventing something being ... what it is'. Hence a 'woman is violated when she finds herself forced to live not as a person but as a mere sex object – a mere sexual resource' (2008: 52). Heidegger would say that, in a similar way, a tree is violated when it reveals itself as mere timber and that a river is violated when it reveals itself to us as nothing more than a source of hydroelectric power. When these things show up for us in these ways they are prevented from being the things that they in fact are. One could say that when the world matters to us in only one way, we find ourselves presented with nothing more than shadows of things, with *Bestand*.

The second point I would like to make is that in his later writings Heidegger insists that 'values-thinking' is inextricably linked to the rise of technology. To think in terms of values, he maintains, is to be blind to the various ways in which things matter to us, to violate them, to rob them of their worth (*BW*: 251).

Φ

So Heidegger clearly thinks that both technology and values-thinking are, broadly speaking, bad things. But what reasons does he provide in support of these conclusions? He certainly does not provide any crisp and clever arguments (by the time he was promulgating his view that values-thinking was a 'blasphemy against Being' he was already well on his way towards his later, notorious view that intelligibility is 'suicide for philosophy' (*CP*: 307)). But that is not to say that his thesis is ungrounded and arbitrary. Instead, as so often in his later works, Heidegger takes a genealogical approach, developing a narrative, the 'history of Being', in the context of which certain philosophical concepts and approaches are revealed as being inadequate. And this is what he does with technology.

Φ

A complete explanation of Heidegger's 'history of Being' would take us too far off course. But here a very brief sketch will suffice.[4] (Since even a sketch of Heidegger's extremely complicated history of Being will be fairly complicated, readers who are either unfamiliar with or allergic to Heidegger's writings might like to skip ahead to Section 3.3.)

The most striking feature of Heidegger's history is its pessimism. Whereas Hegel (ever the enlightenment thinker) thought the history of the world was essentially one of progress, Heidegger (ever the anti-modernist) is far less optimistic. He insists that the history of the West, from the ancient Greeks to the present age, is a story of ontological degeneration, marked, on the one hand, by the gradual 'forgetting' of Being and, on the other, by the rise of an impoverished way of understanding the world.

Heidegger refers to this impoverished understanding as 'metaphysics' (here used as a term of art), and he sees its origins in the Early Greek (i.e., Presocratic) understanding of Being as *physis*. While we moderns, if we think on the matter at all, tend to associate the word with the subject matter of a specific field of scientific inquiry, Heidegger maintains that the Early Greeks saw *physis* as governing the coming-to-be, the 'self-blossoming emergence', of all beings – sky, earth, plants, animals, even humans and gods (*IM*: 14). He expresses this idea by writing that for Heraclitus et al., *physis* governed not just what we nowadays call nature (i.e., the physical realm), but the entire realm of truth. Truth is here being understood in an idiosyncratic way as neither correspondence with reality nor coherence with other truth claims, but as unconcealment (*aletheia*). This is not the place for a thorough explanation of Heidegger's position on this topic; here it is enough to note that he takes the privative alpha- prefix of the term *a-letheia* to convey the idea that *aletheia*, as *un*concealment, depends on a dimension of 'concealment', such that truth, in this sense, means 'what has been wrested from hiddeness' (*Pa*: 171). Now although the Early Greeks experienced unconcealment (i.e., the 'self-blossoming emergence' of *physis*), Heidegger maintains that they failed fully to appreciate the concealed dimension on which *physis/aletheia* rests. In other words, they failed fully to appreciate what it means for anything to reveal itself to us as a being; they failed to appreciate Being. As such, Heidegger concludes, their meditations on *physis* prepared the ground for the gradual alienation from Being which has marked Western metaphysics ever since.

If the Early Greeks prepared the ground for the advent of metaphysics, Plato was its first exponent, the first metaphysician. In support of this claim, Heidegger points out that Plato's idea, as self-present or self-manifesting, does not depend for its disclosure on anything. Reiterating Plato's visual imagery, he claims that the idea, as pure shining, 'does not first let something else (behind it) "shine in its appearance" ... it itself is what shines, it is concerned only with the shining of itself' (*Pa*: 173). But to say this is, once again, to overlook the dimension of concealment that provides the context for all disclosure. In other words, that which grants all beings their being, the 'idea of ideas', is understood by Plato to be 'that which most shines (the most able to shine) *of beings*' (*Pa*: 175, my emphasis).

In Heidegger's view, this is thoroughly misguided. However it is to be understood, Being cannot represented, for to represent it is to suppose that it is a being and to do this is to forget the crucial 'ontological difference' between Being and beings. Indeed, Heidegger sees in the urge to represent Being an urge to dominate, a drive, as it were, to capture Being in the understanding. 'Representing', he writes, 'is making-stand-over-against, an objectifying that goes forward and masters'; it is a 'laying hold and grasping of'; in it, 'assault rules' (*QCT*: 149–50). And Heidegger links this urge to dominate with a tendency to hypostatise the subject. Thus he argues that, in laying the grounds for the metaphysical project of representing Being, Plato set Western thought on a trend towards 'subjectivism' (*QCT*: 128). Heidegger's history of Western metaphysics is therefore also a history of subjectivism. Or, more precisely, it is a history of the *rise* of subjectivism. As one could argue that Descartes, in his securing of the subject in the certainty of its representations, was drawing out the subjectivist consequences of Plato's original account, so one could argue that Fichte's idea of the self-positing Self, for instance, was a product of Kant's desire for an unrestrained noumenal will. Continuing interpretations of this kind, Heidegger sees in Nietzsche, and especially in his notion of the will to power, a distillation of the subjectivism inherent in metaphysics. This accounts for his otherwise startling assertion that Nietzsche represents the last metaphysician and indeed 'the most unrestrained Platonist in the history of Western metaphysics' (*Pa*: 174). However, although Nietzsche represents the final chapter in metaphysics, Heidegger claims that the subjectivist spirit to which he gave voice lives on in the naked will to will which drives technology.

Nevertheless, although metaphysics was misguided, it took its inspiration from the question of Being itself. Now, in the wake of the death of God and in the midst of technology, Heidegger claims that we find ourselves in an era when the question of Being can no longer even be raised. We live in what Michel Haar has called an 'oblivion of oblivion', oblivious not only to the question of the meaning of Being, but also to the fact that we are so oblivious (1987: 87). This is so because technological culture has been largely divested of the representations of Being devised by the metaphysicians of earlier eras. Spinoza's Substance or the Judaeo-Christian creator God were, to be sure, misguided attempts to *represent* Being, but even so they were attempts to represent *Being*, their authors were at least struggling to make sense of Being, however misguidedly. In the technological age such attempts have become increasingly rare, since technology, as Haar puts it, 'hinders an interrogative approach into its own sense'

(1987: 87). 'We must presume', Heidegger writes in his *Nietzsche* lectures, 'that philosophy will disappear as a doctrine and a construct of culture...' (N3: 250). After all, pragmatic technological man knows that God is Dead (although Nietzscheans would contend that he still falls short of an authentic acceptance of this). He knows, moreover, that philosophy has value only to the extent that it is useful – as a source of transferable skills, perhaps, or as a 'training for the mind', or something of that order.

Φ

In this fashion, Heidegger maintains that Plato set Western thought on the road that would lead to metaphysics and eventually to technological nihilism. But how does all this relate to value and value-thinking?

Heidegger's thoughts on this matter are hard to decipher. But, so far as I can make out, his account seems to run something like this. While, over the course of history, metaphysicians found themselves increasingly prone to trade in representations of Being, they found themselves more and more distanced from that mysterious self-concealing dimension – 'Being as such' or *Seyn* – on which those representations depend. Hence the rise of metaphysics involves a gradual alienation from Being or, in less anthropocentric terms, Being's gradual 'withdrawal'. Now Heidegger implies that, over time, metaphysical thinkers came to sense that their attempts to represent Being – as Idea or Substance, or whatever – were in some way inadequate. And he maintains that moral philosophy developed as a direct response to this sense of inadequacy. To 'round out' their unsatisfying representations of Being, metaphysical thinkers postulated what Heidegger calls 'the ought', something that '[B]eing never is yet but always *ought* to be' (*IM*: 197; cf. *BT*: 133). 'The ought', moreover, was thought to have its 'ground in itself', that is, not in Being, but in something 'which in itself raise[s] a moral claim, which ha[s] an intrinsic *value*' (*IM*: 198). So now, with the advent of values-thinking, metaphysicians found themselves, not with a rich account of Being, one that preserves its self-concealing dimension, but with an insipid conception of Being (materialism, perhaps) with certain values, as it were, stuck on (cf. *O*: 68). Indeed the values were added precisely in order to compensate for the impoverished understanding of Being. As Heidegger puts it:

Where anything that is has become the object of representing, it first incurs in a certain manner a loss of Being. This loss is adequately perceived, if but vaguely and unclearly, and is compensated for with corresponding swiftness through the fact that we impart value to the object ...

(*QCT*: 142)

Furthermore, just as the urge to represent Being, to make it entirely accessible to the understanding, may be considered the manifestation of a drive to control and dominate beings, so Heidegger maintains that a tendency to think in terms of values may be associated with a drive to make all things available to the understanding and thus amenable to manipulation and control (see Emad 1977: 201). For this reason, 'thinking in terms of values [as in Nietzsche's work, for instance] contributes most vigorously to the oblivion of Being' (Emad 1977: 201). And for this reason, too, values-thinking is thought to go hand in hand with the naked will to will that lies at the heart of technology.

Φ

Abstruse though they are, Heidegger's claims chime with some of the ways we think and talk about values – especially our tendency to think and speak of the values that things *have*, but also, perhaps, our tendency to refer to things as *being* valuable. Consider the pervasive rhetoric of 'value added', for instance. Nowadays, in considering some proposal, it is fashionable to ask where the value is added in the proposal in question. So, for example, in considering a professor's proposal for an undergraduate course on some technical topic, an administrator might ask where in his course value is added. The demand is that whatever benefit the students might accrue from taking the course be made explicit and available, and preferably in a way that is readily understandable to bureaucrats who know nothing at all about teaching, let alone the topic in question. Ideally, the demand is for something quantifiable, and here talk of value, derived as it is from the language of economics, is particularly useful (see *QCT*: 71). But for Heidegger, these strategies are thoroughly misguided. Just as the essence of good teaching cannot be expressed in a list of figures or bullet-pointed items, so the mystery of Being cannot be conveyed through talk of value.

So Heidegger would have abhorred the rhetoric of 'value-added'. Moreover, just as he deplored technological conceptions of nature, so, deeming all talk of values technological, he would have reserved a special distaste for talk of *environmental* values. Consider, for example, the practice among environmental economists of regarding the total value of a natural 'resource' as the sum of its use values, indirect services and existence values. On this view, the total value of the resource is cashed out in terms of the different kinds of value it possesses. I say different *kinds* of value, but it should be noted that these different kinds are all thought to be commensurable; indeed they are all expressed in monetary terms (more precisely, the amounts the subjects affected would be willing to pay for the resource in question). In this fashion the consumptive use value of, for instance, a stretch of woodland – its use as a source of timber, say – can be given a cash value and directly compared with its existence value, the satisfaction we derive simply from knowing that it exists. Furthermore, the total value of the wood can be directly compared with the total value of some other resource – a bypass, perhaps, or a shopping mall. The one can be weighed against the other, and the most value-efficient course of action determined. The economist is ready to consider any of the manifold ways in which nature matters to us, but only if they can be expressed in pounds and pence or dollars and cents. One can well imagine Heidegger's disgust.

For Heidegger, then, to think in terms of values is to be blind to the various ways in which the natural world matters to us. And so, even against the wishes of those who employ them, references to environmental values reflect and foster an exploitative, instrumentalist stance towards the natural world.

Φ

Of course Heidegger might be wrong about all this, and indeed his case is open to several forceful objections. First, Heidegger's later criticisms of values-thinking are based on his extremely bold account of technology, which is open to challenge. (It is easy to develop Heideggerian readings of putatively technological practices; more difficult to demonstrate the truth of Heidegger's claims.) Second, although *some* references to environmental values seem to evince a technological understanding of things, a host of others do not. (A number of works could be cited here: see, for instance, the impressive non-technological understanding of value set out in O'Neill et al. 2008.)

Heidegger's net catches some fish, but not all. Just as the Heidegger of *Being and Time* was wrong to suppose that all values-thinking presupposes the notion that the world is at root an objective realm standing over against us subjects, so the Heidegger of 'Letter on humanism' was wrong to suggest that all values-thinking is inherently technological. And the same holds true of the criticisms levelled at talk of values by contemporary Heideggerians such as Bruce Foltz (e.g., Foltz 1995: 171). Just as neither Heidegger nor his latter-day defenders have good reason to conclude that there is something objectionable about values-thinking as such, so we should not feel compelled to give up on all talk of environmental values. On the contrary, there are different kinds of values-thinking, 'good' and 'bad', and although Heidegger's analysis can help ethicists to sort the one from the other, it cannot be used to undercut both.

3.3   The hegemony of values-thinking

Yet although there is nothing inherently pernicious about references to environmental values there is, I suggest, something problematic about the notion that *all* the various ways in which nature matters to us can be framed in such terms. Values-thinking is inherently *limited*, even if it is not inherently *pernicious*.

I will argue for this conclusion below. However, it might seem that any argument I might proffer will not be able to get off the ground. For although nature matters to us in a wide variety of ways, it might seem that that variety could always, in principle, be conveyed by postulating different values and different kinds of value. If a man appreciates the beauty of a particular natural setting, then one could say that he is seeing aesthetic value in it. If a woman regards nature as a source of spiritual lessons, then she is seeing spiritual value in it; and so on. Value monism may be a bad thing, but the antidote is to be found, not in the abandonment of values-thinking *per se*, but in value pluralism.

There is something to this. If we are to do justice to nature's value, then we should not rest content with bland references to the value of nature, as if that value were all of one kind. Instead we should investigate the many different kinds of value that nature has and the variety of responses those values can elicit (see further, Embree 1997).

Yet such investigations can only take one so far. For it would be a mistake to suppose that all the ways nature matters to us in the living of our lives can (or could) be understood in terms of environmental values. And the reason for this, I would add, is not that there are too many values, and too many kinds of value, in nature to understand. One cannot adequately understand all the ways nature matters to us in terms of value because many of those 'matterings' are best conceived not in terms of value but in other ways.

Φ

I am not the first to have suggested something along these lines. Christine Swanton, for instance, has argued that many of our moral responses to the world are not best understood in terms of value. Take grief, for example: 'The appropriateness of grief as a virtuous moral response', she writes, 'is not characteristically dependent on the degree of *value* of the item grieved for, but on the *bonds* between the person grieving and the person grieved for' (2003: 42; my emphasis). For this reason alone (though she does provide others) we should be suspicious of the supposed 'truism' that 'the only "rightness relevant respects" which serve to make one option better than another are the degree of strength of the values which inhere in those options' (2003: 34; cf. 48).

The hegemony of values-thinking has also been challenged by environmental philosophers. Alan Holland, for example, has suggested that 'environmental decision-making should be concerned with the continuation of meaning rather than the preservation of value' (2007: 1; cf. Evernden 1985: 124). He suggests that in many cases one can judge the moral rightness or wrongness of a particular action by assessing how it affects the meaning of a particular environment.[6] Hence some actions are morally wrong, not exactly because they reduce the amount of value in the world, but because they irrevocably alter the meaning of a particular environment.

I think Swanton and Holland are on the right tracks here. It seems to me that many of our moral relations with the natural world are simply not best understood – or perhaps, not understood at all – in terms of value.

But in order to show this, more needs to be said. In the following, then, I examine three passages, each of which illustrates a familiar moral relation with the natural world, but one that cannot, I suggest, be adequately comprehended in terms of value.

Example no. 1 (Alice Walker)

> I cried one day as I talked to a friend about a tree I loved as a child. A tree that had sheltered my father on his long cold walk to school each morning: it was midway between his house and the school and because there was a large cavity in its trunk, a fire could be made inside it. During my childhood, in a tiny, overcrowded house in a tiny dell below it, I looked up at it frequently and felt reassured by its age, its generosity despite its years of brutalization (the fires, I knew, had to hurt), and its tall, old-growth pine nobility. When it was struck by lightning and killed, and then was cut down and made into firewood, I grieved as if it had been a person.
>
> (1994: 95)

To be sure, the passage will not appeal to everyone. To some it will seem cloyingly sentimental, and perhaps disingenuous (are we to believe that Walker really grieved for the tree as if for a person?). Nonetheless, the passage gives voice, albeit in an exaggerated form, to a familiar kind of moral response to the natural world.

But how are we to understand that response? Walker clearly valued the tree; perhaps she valued it in a number of different ways, for a number of different reasons. But more than this, she writes that she *loved* the tree. In declaring this was she merely saying that she saw a particular kind of value in the tree? The point is moot.[7]

It is clear, however, that Walker *grieved* for the tree. And with Swanton, I'd want to say that that grief must be understood, not merely with reference to the value she found in the old pine, but to the bond she felt for it. To overlook her bond with the tree, and to try to understand the moral significance of the passage in terms of value and nothing else, is to proffer a reading that is at best incomplete, at worst distorted.

Example no. 2 (Henry David Thoreau)

> We need the tonic of wildness, – to wade sometimes in marshes where the bittern and the meadow-hen lurk, to hear the booming

of the snipe; to smell the whispering sedge where only some wilder and more solitary fowl builds her nest, and the mink crawls with its belly close to the ground. At the same time that we are earnest to explore and learn all things, we require that all things be mysterious and unexplorable, that land and sea be infinitely wild, unsurveyed and unfathomed by us because unfathomable. ... We need to witness our own limits transgressed, and some life pasturing freely where we never wander.

(1999: 282–3)

Thoreau does not say that we value wildness: indeed he laments the fact that, for the most part, we do not value it at all. Nor does he say that we ought to value wildness, though he would not have denied the truth of this claim. Instead, he foregoes the language of value and writes rather of *need*. I suspect that he chooses to do this because, regardless of the intentions of the speaker, talk of values lends itself to being cashed out in terms of preferences and desires. To speak of what an individual values is not usually to say something about that individual – about who she is – but merely to convey what she prefers. But talk of needs runs deeper (cf. Swanton 2003: 47). It is not so much that we value food and water. We need these things.[8] Likewise, Thoreau writes that we *need* wildness, that we *require* it. To be sure, we do not need it to survive. Yet sitting out by Walden Pond, Thoreau has come to realise that he would not be the person he is were it not for his involvement with wild nature. More generally, he maintains that none of us would be the people we are were it not for the comforting and humbling sense that there are still wild places and wild things in the world. The need here must be understood with reference to our second (learned) nature rather than our first (biological) nature; yet for all that, it is still a need.

Example no. 3 (Aldo Leopold)

[O]ur appreciation of the crane grows with the slow unravelling of earthly history. His tribe, we now know, stems out of the remote Eocene. The other members of the fauna in which he originated are long since entombed within the hills. When we hear his call we hear no mere bird. We hear the trumpet in the orchestra of evolution. He is the symbol of our untameable past, of that incredible sweep of millennia which underlies and conditions the daily affairs of birds and men.

And so they live and have their being – these cranes – not in the constricted present, but in the wider reaches of evolutionary time. Their annual return is the ticking of the geologic clock. Upon the place of their return they confer a peculiar distinction. Amid the endless mediocrity of the commonplace, a crane marsh holds a paleontological patent of nobility, won in the march of aeons, and revocable only by shotgun. The sadness discernable in some marshes arises, perhaps, from their once having harbored cranes. Now they stand humbled, adrift in history.

(1968: 96–7)

What, in Leopold's view, is the moral significance of cranes? He clearly believes that their presence makes the world a better place. And he clearly thinks that we ought to ensure their continued survival. Yet he does not maintain that cranes possess some property that confers moral status on them. He does not claim that we ought to keep them around because they are sentient beings or teleological centres of life. Instead, he maintains that whatever worth the cranes have is at least in part a function of their peculiarly intimate connection with the places they inhabit.

This connection should not be conceived in naturalistic terms. Leopold is not suggesting that the cranes have an important *causal* influence on the marsh as a whole. So, for instance, he is not claiming that the crane is what ecologists call a 'keystone species' (though it might well be).[9] Instead, he is making a point about meaning. The cranes, he writes, are a 'symbol of our untameable past', such that when they depart from a particular marsh, the meaning of the place changes. It becomes shot through with a certain kind of sadness. In a Heideggerian fashion, one could say that, by virtue of the special meaning they embody, the cranes serve to 'gather' the world of the marsh.[10]

True, all this *could* be forced into the idiom of value. One could say that the cranes have a certain kind of moral value because of the way they symbolise the marshes. But it seems more natural simply to speak in terms of *meaning*. The cranes have a certain kind of symbolic meaning, and to a large extent this is why they matter, and why, moreover, they matter morally. And, one might add, it is why they *should* matter morally.

Of course many questions remain. What kinds of meaning does Leopold find in the cranes and in the marsh? Which of these meanings can be associated with treating the birds or their habitat well? For present purposes, however, it is enough to note three things: (1) that the cranes (and the marsh) clearly *matter* to Leopold; (2) that this mattering is *morally* significant; and (3) that it is most naturally expressed in terms of *meaning*.

Φ

We have been trying to understand the moral significance of the natural world. And to do this we have examined the testimonies of three individuals for whom nature matters, and for whom, moreover, it matters in a morally significant way. In each case, we have found ourselves presented with a set of moral responses that cannot adequately be framed in the idiom of what Heidegger calls 'values-thinking'. It is true that Walker, Thoreau and Leopold all value nature. Yet to try to capture everything they have to say in the idiom of value is, in each case, to proffer an interpretation that is forced or violent. It is to fail to let the testimony speak for itself.

And I would go further. If it is coercive to impose the single interpretative framework of values-thinking onto the writings of Walker, Thoreau and Leopold, then it is also coercive – even more so – to maintain that *all* the many and various ways in which nature matters to us can be cashed out in terms of environmental values. It is not only that to try to capture everything that has been *written* about how nature matters to us in terms of value is to miss a great deal. It is not just that to try to *conceive* all those different 'matterings' in terms of value is to fail to appreciate many of them. The real danger, I believe, is that a preoccupation with values-thinking might lead one to suppose that it is possible to distil the value from a thing and in this way capture all the ways it matters to us. For having distilled the value of the thing, one might be led to think that the thing itself merits no further attention. Everything that is important, morally, aesthetically and spiritually, will seem to be encapsulated in a neat package of values. And this, surely, would be the world of lived experience well lost.

Φ

Not everyone will be convinced by this, of course. After all (the critic will respond) one can always say that one values X *because* of one's bonds with it, or *because* of one's needs, or *because* of the meaning it has. Nature matters to us in many different ways, to be sure, but the idiom of value is able to capture them all.

But this fails to convince. Once one's bond with a particular natural place has been revealed, what does it add to speak of how one values that place? If one is aware that one needs a particular thing, why add that one values it? If some natural phenomenon has a certain meaning, why then should one feel required to refer to its value? All the conceptual work has been done by the references to bonds, needs and meaning. Once one has explained one's bonds with a thing, one's need for it or the meaning it has, there may be nothing more to say, and nothing more that *should* be said.

I do not mean to suggest that talk of values is always a waste of time, only that it cannot adequately capture all the different ways in which nature matters to us. In many (but not all) cases, that mattering can best be understood with reference to other factors, and values-talk is simply unnecessary. Accordingly, a myopic focus on values can lead one to overlook the rich variety of ways that nature matters to us. That mattering is rich and colourful. In trying to capture it, we need to work with a rich conceptual palette.

Φ

The form of pluralism I am advocating here is not just a moral pluralism (I am equally suspicious of a one-sided emphasis on aesthetic values); however, I am suggesting that in order to understand the various ways nature matters to us *morally*, we need to work, not merely with a host of incommensurable values, but with a variety of different concepts, including certain kinds of bond, need and meaning.

But which factors will be relevant to environmental ethics? Here is a *very* quick sketch:

First, one will have to consider what sort of object one is dealing with: (1) persons (e.g., Julie), (2) sentient non-persons (e.g., Julie's canary), (3) insentient life (e.g., Julie's pot plant), (4) non-living individual beings (e.g., Julie's teapot), (5) abstract entities (e.g., the species represented by Julie's canary) or (6) places (e.g., Julie's village). Different moral considerations will be relevant in each case. So questions about paternalism will only be relevant if one is dealing with persons, that is, autonomous beings capable of making their own conscious decisions. Questions of whether one is harming or injuring a being, rather than just damaging it, will not be relevant if one is dealing with non-living individual beings, such as teapots.

In determining one's moral relations to these various objects, other considerations will be relevant. Often, time-dependent factors will need to be considered. So, for example, in determining one's moral obligations with respect to certain types of object (corpses, heirlooms, particular places, etc.) one will need to take into account certain 'backward-looking' considerations (whose body one is dealing with, perhaps, or the history of the place in question). In other cases (e.g., the treatment of foetuses), the relevant considerations will be 'forward-looking', functions of the being's projected future.

In addition to this, one's moral obligations with respect to the various types of objects listed above will be affected by certain facts about the agent whose obligations are being considered. So in some cases it may be relevant that one has a special bond with the being in question – a personal relationship, say, or even a duty of some other non-personal sort. Or one may need to consider the agent's personal circumstances. If, for instance, the agent has recently suffered a bereavement, then she ought not to be castigated for reneging on her promise to accompany a friend to a dinner party.

Furthermore, there are good reasons to think that in reflecting on what we ought to do we shouldn't just be concerned with our obligations to treat certain beings in certain ways. For one thing, there is more to ethics than matters of obligation. If Bernard Williams is correct (and on this point I think he is), such talk reflects a peculiarly narrow view of what it means to live a worthwhile life (see Williams 1985: Chapter 10). For another, there is no good reason to think that morality must always be understood in terms of the effects one has on the world. As Iris Murdoch (2007) has argued, there is such a thing as 'moral vision' (of which, more in the next chapter). Moreover, in some cases, an action counts as morally right, not because it brings about a change in the world, but because it expresses (or honours) a particular ideal (see further, Swanton 2003: Chapter 6).

Φ

Two caveats. First, and as I said, that was a very quick sketch of the various ways in which nature matters to us, morally. I do not claim that it is exhaustive. Second, I must stress that my aim in distinguishing these various moral considerations is not primarily to categorise them, to chart the moral terrain, as it were, but rather to convey their sheer variety, and by so doing to foster a sense of receptivity or openness in moral deliberation. As Louis MacNeice (1979: 30) wrote, 'World is crazier and more of it than we think / Incorrigibly plural'. This is a salutary reminder, not just when we are investigating what *is*, but also when we are considering what *ought* to be.

No doubt some will want to complain that such efforts to register the richness of our moral lives are at odds with the demand to *act*. After all, it may very well be the case that the moral life is rich and complicated, but if moral philosophy is to have any fairly immediate and tangible effect on the world, then it will have to yield results that are simple enough to be put into practice. In particular, how, if we are to resist the hegemony of value, are we to decide on environmental policy? If all our concerns are pitched in terms of value and its promotion, then our way would seem to be clear: policymaking can involve the weighing up of values, of pros and cons, and the rules for doing this can in principle be followed by all, thus ensuring fairness.

But what if we are to consider not simply values, but how a proposed course of action fosters certain *bonds* or satisfies certain *needs* or preserves certain kinds of *meaning*? How then are we to decide? For instance, can the promotion of certain values be weighed up against the preservation of certain meanings?

It is admittedly difficult to see how talk of bonds, needs and meaning might be amenable to the decision procedures currently in vogue among environmental policymakers. But this may not prove an insurmountable problem. After all, decision-making need not involve what Heidegger called 'calculative thinking', the mechanical totting up of pros and cons. Decisions can be made, and policy decided upon, in other ways too. And there are other models to choose from here – legalistic ones, for example. Suffice it to say that there is no reason why resisting the hegemony of value should leave policymakers all at sea.

Notes

1. This has implications for how one is to understand the reality of nature, conceived as a realm independent of human concerns. I discuss these issues in Chapter 5.
2. Geoffrey Lipman, former President of the World Travel and Tourism Council, quoted in Neale 1998: 34.
3. Heidegger tended to use the terms 'aesthetic' and 'moral' pejoratively, regarding both as symptomatic of our estrangement from Being. Rather than translating references to things of beauty and what we are obliged to do into clunky Heideggerese, I have chosen to stick with the more familiar terms.
4. For a more thorough exposition, see Thomson 2005: Chapter 2.
5. For persuasive defences of Heidegger's account of technology, see Thomson 2005 and Young 2008.
6. It should be noted that Holland's claims are directed primarily at philosophers. For while, in philosophical circles, values and meanings are often treated differently, in disciplines such as anthropology the two concepts are often used interchangeably.
7. Swanton, for her part, argues that love must be understood with reference to bonds, and not merely to value (2003: 117).
8. In arguing for the existence of objective values in nature, Holmes Rolston III (2003) draws attention to the fact that like other organisms, humans included, plants value things such as sunlight, water and nutrients. But this is an odd claim. It is rather the case that plants *need* sunlight, water and nutrients. And our moral obligations with respect to plants will reflect, not the importance of those needs for the plant, but the importance of plants for us. So if we value some sort of plant, that will provide us with one reason for ensuring that the needs of those plants are met. But if, by contrast, we do not value them – if, for instance, we consider them weeds – then we may have no reason to ensure that those needs are satisfied.
9. A 'keystone' species is one whose effect on the integrity of the ecological communities in which it is represented is disproportionately large relative to the abundance of its members. The loss of the members of such a species from a particular ecological community therefore undermines the integrity of the community as a whole (see further, Smith and Smith 2001: 393).
10. The allusion is to Heidegger's account of 'things', as presented in essays such as 'The Thing' (*PLT*: 165–86) and 'Building Dwelling Thinking' (*BW*: 143–212). In these essays, Heidegger maintains that a thing is a 'gathering' of four dimensions of lived experience: earth, sky, mortals and gods. To be sure, he tends to focus on artefacts (a jug in 'The Thing', a bridge in 'Building Dwelling Thinking'), but he admits that animals can sometimes act as things too (cf. *PLT*: 182: 'Things, each thinging from time to time in its own way', he writes, 'are heron and roe, deer, horse and bull'). So perhaps the cranes could serve to gather the world of the marsh.

4
Why Conserve Nature?

4.1 Phenomenology and moral normativity

In Chapter 3, we saw that our moral relations with the natural world are too diverse to be captured in the monochromatic idiom of values-thinking. To understand those relations we need to consider a range of factors: not just how much value our actions promote but our bonds with non-human others, for example, or the ideals we express through acting in certain ways towards natural things.

By illuminating the diversity of moral considerations, phenomenological inquiries can support a certain kind of moral pluralism. But can any sort of phenomenological inquiry recommend a particular course of action? Can phenomenology generate moral norms? Some writers contend that it cannot. Peter Heath, for instance, writes that 'Phenomenology may ... be useful in studying moral experience, but can do nothing to recommend or vindicate any particular form of morality' (1975: 159). The findings of phenomenology, he adds,

> will be neutral ... in the sense that they cannot and should not be expected to exert any direct influence upon the desires, wishes, feelings, etc., of the phenomenologist himself, or of those he seeks to instruct. [The phenomenologist's] role is merely that of an observer, reporter, analyst, taxonomist and so on, and his sole business is to provide information. Even if, as is at least conceivable, his investigation should reveal 'norms' or 'values'..., they would strictly be 'value-facts' in this context, and his account of

them would likewise be purely descriptive and not normative in character.

(1975: 161)

If Heath is correct, then phenomenological inquiries will not be able to provide any guidance on what we ought to do. And so they will be of no use in helping us decide how we should act in relation to the natural world.

The primary aim of this chapter is to show, against Heath, that phenomenological inquiries are able to generate moral norms that can be fruitfully brought to bear upon the question of how we ought to relate to the natural world. More precisely, I will use a phenomenological approach to demonstrate that it is a good thing – *morally* a good thing – to attend carefully and selflessly to natural things, processes and places. And I will also show, by means of a broadly phenomenological mode of inquiry, that there is a moral imperative to conserve nature.

Φ

The first thing to note, in response to Heath's charge, is that even if phenomenological investigations are inherently descriptive, the descriptions thus produced might nonetheless have implications for normative ethics.

An example should make this clearer. Phenomenologists maintain that in order adequately to describe what it is like to be human, one must distinguish one's experience of objects from one's experience of persons. So, for instance, Husserl spills a great deal of ink describing what it is like to relate to a being as to another person. Likewise, as we saw in Chapter 2, Heidegger explains what it is like to share a world with beings who are in certain essential respects like oneself. Yet whether it is framed in terms of 'persons' or 'Dasein', the basic point is the same: autonomous beings like you and I are experienced as special kinds of being, essentially unlike chairs and tables.

Now, to be sure, all this *is* descriptive: the phenomenologist is describing what it is like to experience beings as persons rather than as things. However, it is important to note that by conveying what it is like to experience another being as a person, the phenomenologist pitches himself against all those who would have us believe that humans are essentially *things* (see further, Ratcliffe 2007). So he stands opposed, not only to all those academic authors who try to explain what it is like to be human in impersonal terms, but also to a whole host of political institutions and socio-linguistic practices. So, for instance, the phenomenologist's account stands opposed to the widespread tendency to refer to 'human resources' (cf. *BW*: 323).

All this holds true of environmental philosophy as well. A phenomenological investigation that brings out the richness of our lived experience of nature should not be dismissed as merely descriptive and bereft of any normative significance. For to convey that richness is to undermine the pervasive notion that nature is nothing more than a sink of resources. The more people read authors like David Abram and Erazim Kohák, the more dissatisfied they will be, not only with those who see nature as nothing more than wasteland and timber, but also with those environmentalists who choose to make their cases in the bland, managerial idiom of natural services, consumptive use values and so on.

It would therefore be wrong to suppose that phenomenology is merely descriptive and without any normative significance. It is true that phenomenology aims at describing the structure of experience, but as Patrick Curry has pointed out, a 'particular description is only one way out of a theoretically unlimited number of ways the world, or part thereof, could be described; it therefore also *prescribes* a particular way of looking at it' (2006: 8). So phenomenology has normative implications in this sense, at least: it prescribes a particular way of seeing things.

But there are other reasons for thinking that, against Heath, phenomenology is not merely descriptive, but normative as well. In order to see them, however, we will have to begin by considering, not environmental ethics, and not even the topic of moral normativity in general. We must begin with the basic question of what it means to do phenomenology.

Φ

As we have seen, phenomenologists typically maintain that we are subject to certain ingrained prejudices which lead us to misconstrue the phenomena (by which I mean not only what presents itself to us in experience but also the manner in which it presents itself). The main culprit here is often thought to be scientific naturalism, a commitment to which is said to encourage the false notion that experience presents us with what Merleau-Ponty calls an an 'objective world', a realm of objects arrayed in Euclidean space and related to one another by relations that are merely external.

Having identified these prejudices (as they regard them), phenomenologists typically proceed to outline some method by which they may be suspended or put out of play. This is not to say that they suspend all belief in a world transcending our consciousness of it: many of their number would reject the early Husserl's contention that such a measure is necessary (or even possible). Yet all phenomenologists would agree that in order to describe the phenomena accurately the phenomena, the phenomenologist must set aside her commitments to scientific or philosophical theories about the nature of things. This is the point Merleau-Ponty is making on the very first page of *Phenomenology of Perception* when he asserts that phenomenology 'tries to give a direct description of our experience as it is, without taking account of its psychological origin and the causal explanations which the scientist, the historian or the sociologist may be able to provide' (*PP*: vii). The point is not merely that such explanations have no place in phenomenological reflection. It is rather that one's commitment to this or that theory could lead one to misconstrue the phenomena. (And it should be noted that the danger here is not only with scientific, historical and sociological theories. Philosophical views can also have a distorting effect.)

To suspend one's theoretical commitments in this way is not to reject them. For instance, following Husserl's approach in the *Crisis*, Merleau-Ponty begins *Phenomenology of Perception* with a *désaveu* of science – not a rejection but a 'foreswearing' of science, in Colin Smith's apt translation (*PP*: viii; see further, Matthews 2002: 32). He is not hostile to science – quite the opposite. Nonetheless, for the purposes of phenomenological reflection, he chooses to suspend his allegiance to scientific theories, so that he might be able to describe the phenomena more accurately.

So to suspend one's allegiance to certain second-order or theoretical commitments is not to reject them. Yet it remains the case that suspending one's commitments in this way tends, as it were, to loosen their hold upon us. For instance, to suspend the assumption that experience presents us with an array of determinate objects arrayed in Euclidean space may not be to reject that supposition outright, yet it is to place it in question and, in this way, to weaken its grip upon us.

Finally, suspending one's theoretical commitments in this way is thought by phenomenologists to afford one a clearer view of the phenomena. One could say that the phenomenologist is released *from* certain views about reality and released *to* the phenomena. Thus Merleau-Ponty maintains that the aims of phenomenology are similar to those of art: both can '[penetrate] to the root of things beneath the imposed order of humanity' (*SNS*: 16) and in so doing bring one 'face to face with the real world' (*PrP*: 159). Both can enable one to 'rediscover phenomena' (*PP*: 57); both can 'teach us to see [things] clearly once more' (*PP*: 456).

Φ

We have been focusing on Merleau-Ponty; however, I think that most (perhaps all) phenomenologists would endorse the following three propositions:

1. We are subject to certain ingrained prejudices which lead us to misconstrue the phenomena.
2. These prejudices can be suspended or put out of play.
3. Doing this affords one a clearer view of the phenomena.

Yet even if all phenomenologists would endorse these views, few of their number would say that this endorsement has any normative significance, morally speaking. Although most phenomenologists would agree that in order to do phenomenology well one ought to suspend certain prejudices, few would say that this 'ought' has any moral force.

In the following, I will argue that, notwithstanding the claims of phenomenologists, the three propositions set out above imply certain moral norms. In other words, I will argue that it is, morally speaking, a bad thing that we are subject to certain ingrained prejudices which lead us to misconstrue the phenomena. And I will also suggest that it is morally praiseworthy, perhaps morally obligatory, to try to suspend these prejudices.

Φ

But before developing my case, I would like to examine in some detail the work of a particular phenomenologist: namely, Heidegger's writings on what he calls 'releasement toward things' (*Gelassenheit zu den Dingen*) (*DT*: 54). This may seem an odd place to look, not least because by the time Heidegger was writing on the topic of releasement he had explicitly distanced himself from the tradition of phenomenology. Nonetheless, I hope to show that his writings on this topic conform to the basic tripartite structure of phenomenological inquiry, as outlined above, and that for this very reason they ought to be regarded as works of phenomenology.

As we have seen, phenomenologists typically maintain that we are subject to certain ingrained prejudices which lead us to misconstrue the phenomena. Heidegger, for his part, maintains that we are subject to 'technology', which, as we saw in Chapter 3, is to say that in the modern age things tend to reveal themselves to us as mere resources (*Bestand*). To be sure, Heidegger rejects the notion that technology is a 'view', a way we interpret the world, still less one that we might be able to relinquish at will. Instead, he insists that it is a 'destining of revealing' (*BW*: 330), by which he means that it marks a turn in the history of Being to which we are subject, rather than a change in our perspective on how things are. The fact remains, however, that like other phenomenologists Heidegger thinks that we are in the thrall of a pervasive understanding (broadly construed) that distorts the way things present themselves to us in experience.

The second general conviction of phenomenologists is that those prejudices to which we are subject can be suspended. Again, but again with certain qualifications, Heidegger espouses a similar view. In his later writings he calls for his readers to exercise 'releasement toward things' and in this way to 'let them be'. To exercise releasement towards things is, he explains, to be released *from* technology. This is not to reject technology, to 'curse it as the work of the devil' (*BW*: 330), but to suspend one's allegiance to it, and so to find oneself brought into a 'free relationship' with it (*BW*: 311). If we exercise releasement, he explains, we will say '"yes" and at the same time "no"' to technology (*DT*: 54) and so 'keep open' to the 'meaning' lying 'hidden' within it (*DT*: 55). In this way we will find ourselves able to 'let technical devices enter our daily life, and at the same time leave them outside' (*DT*: 54).

Now as we have noted, by the time he was writing about technology, Heidegger had explicitly distanced himself from the tradition of phenomenology. Nonetheless, what we have here is a suspension or foreswearing of technology formally similar to the suspension that is an integral part of more conventional varieties of phenomenological inquiry. Just as the phenomenologist suspends her allegiance to certain theoretical views about how experience may be explained, so the 'released' individual suspends his allegiance to technology.

The phenomenologist's third contention is that suspending one's prejudices can afford one a clearer view of the phenomena. Again, the parallels with Heidegger's thought are striking. As we saw in Chapter 3, Heidegger maintains that technology does violence to things, not by literally destroying them as a bulldozer, say, might destroy a meadow, but by preventing them from disclosing themselves as the things they are. To be released *from* the hegemony of technology is therefore to be released *to* things, to find oneself privy to how they really are. Heidegger expresses the point by saying that the released individual finds himself able to 'let beings be', able, that is, to allow things to disclose themselves under their own terms, according to their own possibilities and limitations. So just as, by tending to the needs of his plants, the gardener allows them to flourish, so the released individual is said to act as a 'shepherd of Being' (*BW*: 245), letting things come into their own, allowing them to blossom not *in* the world, but in*to* the world (of which, more presently).

Φ

Heidegger evidently regrets the fact that the world has become technological, and he clearly believes that it is, broadly speaking, a good thing to exercise releasement. Yet he would have rejected any suggestion that we are *morally* obliged to exercise releasement and he would have rejected the claim that releasement is a *moral* virtue.

His stance on this issue reflects his general aversion to ethics. As we saw in Chapter 3, Heidegger came to conclude that the advent of ethics – the establishment of 'the ought', as he puts it – marked a lamentable downturn in the history of Being, a symptom of Being's 'withdrawal'. But it would, I think, be an error to take Heidegger's disavowal of ethics at face value. For his target seems to be, not ethics *per se*, but a particular conception of ethics. His objections seem to be directed towards something close to the conception Bernard Williams once called the 'peculiar institution' of *morality*, that account of the ends of human life, principally associated with Kant, which sees moral concerns as being sharply distinguished from other practical considerations (see Williams 1985: Chapter 10). Heidegger's criticisms do not seem to be aimed at the rather broader concerns of *ethics*, conceived as an inquiry into what it might mean for humans to live lives that are generally admirable and not simply worthy of moral praise. Indeed it seems very likely that, despite Heidegger's claims to the contrary, his writings on releasement could plausibly be framed as a virtue ethics of a roughly Aristotelian kind (cf. Polt 1999: 170). After all, like Aristotle, Heidegger believes that humans have an essence, and that 'doing well' consists in acting in accordance with this essence. To be human, he claims, is to be a space or clearing (*Lichtung*) in which beings come to be, and to live well as a human is to have developed a capacity to 'shepherd' the emergence of these beings. What is more, Heidegger seems to share Aristotle's view that the good life can (for the most part at least) be cashed out in terms of the possession of certain character traits. For Heidegger, such a life would be marked by the variety of selfless attention he refers to as releasement. And it would, one imagines, be free of certain vices – curiosity and distractedness, for instance, or the relentless drive to consume that marks *Homo technologicus*.

Φ

Let us grant that releasement may be considered a virtue in this sense, an integral component of 'the good life' or 'well-being', as Heidegger conceives it. What reasons are there for thinking that it is a *moral* virtue?

For some, the question is redundant, for to show that a particular character trait is an integral component of the good life simply is to show that it is a moral virtue. But let us assume, for the sake of argument, that this is not the case, and that character traits integral to the good life (i.e., virtues) come in both moral and non-moral varieties. Given this assumption, what reasons are there for thinking that releasement is a *moral* virtue? To answer this question we will have to leave Heidegger for a moment and turn to the work of another thinker, namely, Iris Murdoch.

Φ

Although, so far as I'm aware, Iris Murdoch never referred to herself as an exponent of phenomenology, she nonetheless had much in common with that tradition. Like Merleau-Ponty, she emphasises that '[i]t is a *task* to come to see the world as it is' (2007: 89). But she adds that this task is in large part a *moral* one. Examining the reasons she gives for this conclusion will help us see why Heidegger's releasement should be considered a moral virtue. More generally, it will illuminate the connections between what it means to do phenomenology and how one ought morally to comport oneself towards the world.

In a well known discussion, Murdoch makes her case by describing a mother's changing attitudes towards her daughter-in-law. At first, she writes, the mother finds the girl 'pert and familiar, insufficiently ceremonious, brusque, sometimes positively rude, always tiresomely juvenile' (2007: 16–7). She does not like the girl's accent; she 'feels that her son has married beneath him' (2007: 17). The mother, however, is not a bad person; on the contrary, she is both well intentioned and self-critical, and she comes to the conclusion that she might be regarding her daughter-in-law unfairly, that her impressions of the girl might be the expressions of snobbishness, perhaps. So although the girl herself remains unchanged, the mother's view of her changes. She is 'discovered to be not vulgar but refreshingly simple, not undignified but spontaneous, not noisy but gay, not tiresomely juvenile but delightfully youthful, and so on' (2007: 17).

The salient points here are as follows. First, the mother's perception of her daughter-in-law is distorted by self-centred preoccupations such as snobbishness. Indeed, Murdoch claims that it is a general truth that '[o]ur minds are continually active, fabricating an anxious, usually self-preoccupied, often falsifying veil which partially conceals the world' (2007: 82). Hence in trying to attend to reality '[t]he difficulty is to keep the attention fixed upon the real situation and to prevent it from returning surreptitiously to the self with consolations of self-pity, resentment, fantasy and despair' (2007: 89). Second, by noticing and suspending her tendency towards snobbishness, the mother was able to open up a space, as it were, in which the girl could reveal herself 'as she really is' (2007: 36). Again, Murdoch sees a general lesson here. In order to see the world truly, we must suspend our ingrained tendency to regard things (and persons) through the distorting lens of self-interest. Since we are 'naturally selfish' (2007: 76), we must 'unself' if we are to see the world clearly (2007: 82). Third, the mother undergoes a *moral* transformation. Her achievement is to some degree moral in character, a result of 'moral imagination and moral effort' (2007: 36), the manifestation of a 'patient, loving regard' (2007: 39). One could say that her initial impressions of the girl were *unfair*, that they did not do *justice* to who she was. And again Murdoch sees a wider truth here. To regard the world through the lens of self-interest is to prevent beings from disclosing themselves as the beings they in fact are. And it is therefore a good thing, morally speaking, to suspend this tendency, to 'unself'. For this reason, Murdoch suggests that it is a general truth that the development of 'attention ... of looking carefully at something and *holding* it before the mind' amounts to 'moral training' (1993: 3). Fourth, the mother's transformation is moral in character, not only because she has learnt to attend to a moral patient (namely her daughter-in-law), but because she has learnt to attend *per se*. In other words, Murdoch is suggesting that, all things considered, it is morally a good thing to pay attention, regardless of what one is paying attention to. Attention is therefore a virtue that can be directed towards many different kinds of objects, not just persons but things as well. And in line with this, Murdoch suggests that if attention is a moral virtue, then it may also be regarded as an aesthetic one: '[v]irtue is *au fond* the same in the artist as in the good man in that it is a selfless attention to nature' (2007: 40).

Φ

The formal similarities with Heidegger's account are striking. Although Murdoch refers to self-interest rather than technology, she is of a piece with Heidegger in thinking that we are subject to an ingrained prejudice which distorts our experience of the world. She also agrees that this prejudice can be suspended and that doing this can enable one to let beings be the beings they in fact are (to use a Heideggerian expression).

But why exactly does Murdoch consider attention a *moral* virtue? Not simply because exercising attention is a particularly effective means by which to become less self-obsessed. On the contrary,

Murdoch seems to believe that a disposition to exercise attention is an integral component of the good life, and not just a means to it. But this is not to say that attention is a virtue merely because it is of some benefit to the virtuous (i.e., attentive) agent. For, Murdoch seems to think that attention is a virtue, not only for 'eudaimonistic' reasons, not only because it contributes to the good life or to human 'well-being', but also for what might be referred to as other-regarding reasons. It is, she implies, a good (i.e., virtuous) thing to exercise attention because beings *deserve* such attention.

This may seem an odd suggestion. If it makes sense to say that a being deserves a certain kind of response from moral agents, then that response is typically thought to have some publicly observable *effect* on the being in question. So, for instance, it makes sense to say that a horse deserves to be well treated, but only because horses can reap the physical and emotional rewards of proper care. However, merely *attending* to a being cannot benefit it in this way. A capacity to attend to a being in a certain way might incline one to treat it well. Indeed a disposition to exercise this sort of attention might go hand in hand with a certain constellation of behaviours: it might manifest itself in action as gentleness or tenderness, for instance (cf. Cooper and James 2005: 101–3; Benso 2000: Chapter 12). But one might nonetheless doubt whether merely attending to a certain object can be of any benefit to the object in question.

But this conclusion would be hasty, for attending to a being can benefit it in a more subtle sense. Recall Murdoch's discussion of the mother and her daughter-in-law. For Murdoch, what makes attention a virtue is not only the fact that it benefits the attentive agent (though it surely does). At least part of what makes it a virtue is the fact that it lets beings be. The mother's transformation was moral in character, not because it involved some publicly observable change in either the mother or the girl, but (at least in part) because it enabled the girl to disclose herself 'as she really is' (2007: 36). The attention she developed may therefore be counted as a moral virtue, not because it brought about some state of affairs in the public world, but because it allowed the girl to disclose herself truly.

Now something similar can, I think, be said of Heidegger's releasement. As we have seen, Heidegger's account implies that a settled disposition to let beings be is an integral component of the good life. Moreover, like Murdoch's attention, this virtue is other-regarding in the special sense that, when exercised, it allows beings to disclose themselves as the beings they in fact are. Once we have relinquished Heidegger's narrow conception of ethics for a more plausible and wider Murdochian one, there is no good reason to deny that in speaking of 'letting beings be' we are referring to what we ought, morally, to do. There is no good reason to deny that Heidegger's releasement is a *moral* virtue.[1]

Φ

Let us pause for a moment to recap the argument so far. I have argued that Heidegger's writings on releasement have the tripartite form characteristic of phenomenological inquiry, and I have suggested that, despite the man's own objections, they should therefore be regarded as phenomenological investigations. I have also argued that releasement is a moral virtue. If I am right on these points, then Heath is wrong to contend that by its very nature phenomenological inquiry cannot generate moral norms. At least one form of phenomenological inquiry – namely the one set out in Heidegger's later writings – can generate such norms. One kind of phenomenological inquiry can show that we ought, ought *morally*, to develop and exercise the distinctive kind of attention Heidegger calls releasement.

Yet even if one grants both that Heidegger's writings on releasement are examples of phenomenological inquiry and that they embody (or at least can serve as the basis for) certain moral norms, it remains to be shown that other, non-Heideggerian forms of phenomenology can also be framed in this expressly moral way. Even if one grants that Heidegger's rather idiosyncratic variety of phenomenology involves the exercise of something like Murdoch's moral virtue of attention, can the same be said of its Husserlian or Merleau-Pontian counterparts?

At first sight, the answer to this question would seem to be yes. For one thing, phenomenologists in general – and not just Heideggerian ones – tend to think that doing phenomenology well requires the exercise of something like Murdochian attention. Moreover, although they are loath to say that we have a moral duty to do so, their writings often imply that exercising such attention is an integral component of the good life. To be sure, these claims are seldom expressed in this explicitly virtue ethical manner. Yet these views are, I believe, evident in the notion, commonly espoused by phenomenologists, that phenomenological reflection can enable a person to free herself from the prejudices of public opinion and the 'inauthentic' modes of existence which they reflect and foster. This is most evident, of course, in the writings of existential phenomenologists such as Sartre, but it is prefigured in Husserl's mature view that phenomenology enables a man to 'shape' himself 'into the true "I", the free autonomous "I" ... striving to be true to himself' (*CES*: 338). The thought here, surely, is that phenomenological attention is an integral component of the good (i.e., authentic) life – that it is, in broadly Aristotelian terms, a virtue.

But this might seem too fast. After all, and as we have seen, Murdoch seems to think of attention as being directed towards items (daughters-in-law, etc.) 'out there', as it were, in the world. However, phenomenologists such as Husserl and Merleau-Ponty are not concerned with things, so much as our experience of them, and more precisely, the basic structures of experience brought into play in their disclosing themselves as they do. As Husserl maintains, to do phenomenology is to 'be on the lookout for things or objects in the life-world, not in order to know them as what they [really] are but rather in order to inquire into the modes of their subjective manners of givenness, i.e., into *how* an object ... exhibits itself as being' (*CES*: 159; editor's annotation).

The worry here is *not* that the phenomenologist is concerned with the thing-as-experienced rather than the thing-in-itself. The worry is rather that the phenomenologist only attends to the thing as experienced in the hope of discovering certain general truths about what it is like to experience anything. The troubling thought, then, is that the aim of phenomenological reflection might not be to let beings be, but rather to illuminate 'their subjective manners of givenness', in Husserl's sense – quite a different matter.

Once again, however, this worry is groundless. It relies on there being a difference in kind between attending to *what* discloses itself in experience (and so remaining within what Husserl calls 'the natural attitude') and attending to *how* objects disclose themselves ('their subjective manners of givenness'). Yet there are good reasons for doubting whether any such clear distinction can be drawn. Merleau-Ponty, for instance, suggests that by attending to what discloses itself, one can bring to light how it discloses itself. Thus he maintains that a great artist such as Cézanne is able, by attending to things (apples and oranges, Mont Sainte-Victoire, etc.), to reveal that which is also revealed by phenomenological inquiries, namely, the pre-objective world. To '[penetrate] to the root of things beneath the imposed order of humanity' (*SNS*: 16), the artist, like the phenomenologist, must comport himself towards things in such a way that he lets them be, allows them to disclose themselves as the things they in fact are. Only through this will he be brought 'face to face with the real world' (*PrP*: 159).

I am suggesting that phenomenologists – and here I mean phenomenologists generally, rather than just Heidegger – exercise something like Murdochian attention because they seek to do justice to the phenomena, where by the phenomena I mean neither things 'in themselves' (as opposed to our experience of them) nor our experience of things (as opposed to how they really are). Now non-Heideggerian phenomenologists do not have a great deal to say about the dispositions of character required to attend to the phenomena. One will search the works of Husserl or Merleau-Ponty in vain for discussions of anything like Heidegger's notion of releasement. Yet their attention to the phenomena is nonetheless expressed in their writings. So whereas Merleau-Ponty, for instance, does not follow Heidegger's lead in writing about releasement, something like that virtue is expressed in his wonderful, evocative descriptions of how things present themselves to us. Merleau-Ponty's works do not make a case for attention, and in this respect they are unlike Murdoch's works or those of Heidegger. Indeed they are not *about* attention at all. Instead, they are examples of attention in practice. As the man himself put it, phenomenology embodies 'the same kind of attentiveness and wonder, the same demand for awareness', that one finds in the works of Balzac, Proust, Valéry and Cézanne (*PP*: xxi). And this attention is there, too, in Merleau-Ponty's own writings, and in the best work of his fellow phenomenologists.

4.2. Selfless attention to nature

If I am right on these points, then Heath is wrong to suggest that phenomenology can yield no moral norms. For, if the case set out above is correct, then the findings of phenomenology both reflect and foster the exercise of a particular moral virtue, namely, attention. They reflect and foster a settled disposition to 'let beings be'.

Now if it is a good thing to develop and exercise the virtue of attention, then it is a good thing to be attentive in one's relations with nature. But this is of course trivially true. Is there a stronger and more interesting connection between the exercise of attention and our relations with nature?

In *The Embers and the Stars*, Erazim Kohák suggests that there is. He begins by noting what many others have noted before him: that materialism fails to do justice to the richness and depth of our lived experience of nature. What, then, in the midst of our modern materialistic world, is one to do? Kohák's first recommendation is unsurprising. The dedicated phenomenologist must 'suspend all theory and ask, without prior ontological prejudice, just what it is that in truth presents itself in lived experience' (1984: 22). But this, Kohák maintains, will not be enough. We modern, city-dwelling phenomenologists will find that even if 'we bracket the concept of "nature" as a mechanical system and of the human as the sole source of meaning, our urban experience will lead us right back to it' (1984: 23). In other words, our lived experience of nature has become so impoverished that it mirrors and so confirms our impoverished conception of what nature is. We have, in effect, 'translated our [materialist] concepts into artifacts' (1984: 12), such that we now find ourselves 'amid a set of artifacts which conform to our construct of reality as matter, dead, meaningless, propelled by blind force' (1984: 12–13). For this reason, Kohák proposes that the traditional Husserlian bracketing be augmented with a 'practical' bracketing, a suspension not just of concepts but of the artefactual world. And to perform this second bracketing, we must literally remove ourselves from urban environments; we must take to the woods.

This is an appealing suggestion (and beautifully expressed by Kohák), but it is in the final analysis unconvincing. The variety of phenomenology Kohák endorses is aimed at revealing what Husserl calls the 'subjective manners of givenness' of things. However, these manners of givenness are so general in character that they will apply regardless of whether one is dealing with natural or non-natural intentional objects. My television set is intersubjectively constituted, yet the same may be said of the horse chestnut I pass each morning on my way to work. Only one of these objects is natural; however, in this context, that fact is irrelevant.

Nonetheless, though the reasoning in support of it is flawed, Kohák's conclusion is, I believe, correct. He is right to suggest that encounters with nature can aid the phenomenologist, not exactly because they can help one discover the fundamental structures of experience, but because they can help one develop the attention which is the mark of phenomenological inquiry done well.

Φ

The connection between one's material circumstances and one's ability to develop and exercise attention is conveyed in the following passage from Nyanaponika Thera:

Apart from (supposedly) disinterested scholarly or scientific research, man is generally more concerned with 'handling' or utilizing things, or defining their relations to himself, than with knowing them in their true nature. He is therefore mostly satisfied with registering the very first signal conveyed to him by an inner or outer perception. Through deeply engrained habit that first signal will evoke standard responses by way of judgements such as good-bad, pleasant-unpleasant, useful-harmful, right-wrong: which again will lead to further reactions by word or deed in accordance with these judgements. It is very rare that attention will dwell upon an object of a common or habitual type, any longer than for receiving that very first, or the first few signals. Thus only one single aspect of the object, or a selected few will be mostly perceived and sometimes misconceived; and only the very first phase or little more of the object's life-span will come into the focus of attention. One may not even be consciously aware that the respective process has an extension in time (origination and end); that it has many aspects and relations beyond those at first sight connected with the casual observer or the limited situation; that, in brief, it has a kind of evanescent individuality of its own. A world that has been perceived in that superficial way will, to that extent, consist of rather shapeless little lumps of experiences marked by a few subjectively selected and sometimes misapplied signs or symbols which have significance mainly for the individual's self-interest.

(1971: 35–6)

Nyanaponika, a Buddhist monk, is discussing our habitual tendency to see the world through the prism of self-centred (or more broadly, anthropocentric) concerns. And in keeping with Buddhist philosophy, he suggests that the key to overcoming this tendency lies in a transformation of mind.[2]

In this sense, then, Nyanaponika focuses on the 'subjective' aspect of what he regards as our habitual lack of attention. It is ultimately up to us whether we attend to things or not. If we are preoccupied with handling and utilising things, then the world will seem to 'consist of rather shapeless little lumps of experiences' (cf. *PP*: 322). But if, by contrast, we develop the virtue of mindfulness or attention we will find ourselves able to appreciate things in their 'evanescent individuality'. Yet it is interesting to note that at one point Nyanaponika shifts emphasis, writing that it will be more difficult to exercise attention when one is dealing with objects of a 'common or habitual type'. Here he is referring, not to the 'subjective' conditions brought into play in the perception of an object, but to the object itself. He is pointing out that certain types of object and certain material conditions encourage our habitual inattentiveness.

The worst offenders here will, I suggest, be certain kinds of mass-produced object. So, to take a concrete example, as I write these words I am sitting opposite my television set. It discloses itself to me as representing a type – it is just one Bush RF2185NTXSIL, identical in almost every respect to the hundreds (perhaps thousands) of Bush RF2185NTXSILs tucked into the corners of living rooms across the country. It does not reveal itself as a thing in its own right (what could be special about *this* television set?) and so it does not reveal itself as worthy of attention. And of course the same could be said of the other mass-produced items that surround me – the sofa on which I am sitting, for instance, or the lamp above my head, or the laptop computer on which I am typing. But objects of this kind fail to invite attention for two reasons; not just because they have been mass-produced, but also because they have been designed to serve certain functions. So, to remain with the example of my Bush RF2185NTXSIL, the sheer functionality of the thing tends to inhibit the exercise of attention. In normal circumstances it would not enter my head to regard it as a thing in its own right – I simply switch it on, park myself in front of it and become absorbed in whatever show has been scheduled. One could say that the television set reveals itself to be little more than a means to some immediate practical end, my viewing of such and such a programme. There seems to be nothing more to it. Even its sleek contours, smooth finish and clean edges reflect a kind of functionalist aesthetic, the longing for efficiency that marks the technological age. Even its aesthetic properties seem to point beyond the thing itself to certain human ends.

Items of this sort, what I shall call 'merely functional' things, disclose themselves, not as things in their own right, but as what Heidegger calls *Bestand*, 'standing reserve'.[3] For any such thing always seems to point beyond itself to some use to which it might be put, some end it might help one procure. In a manner of speaking, the merely functional thing has become reflective, like a mirror. For in looking at it there is nothing to arrest one's gaze, and so one sees nothing more than a reflection of certain human concerns – this or that project, this or that purpose, this or that state of affairs one desires to bring about.

Accordingly, a *world* in which all things had come to disclose themselves as merely functional would be one in which one's surroundings had become reflective, one in which men, women and children could see nothing but themselves, as Heidegger might have said (cf. *BW*: 332; cf. Abram 1996: 22). It is easy to imagine such a world – after all, it would not be so different from the one we currently inhabit. It would be a world in which things are specifically designed not to hold attention, but instead to reflect our cravings for sugar, caffeine, alcohol or nicotine, for power, status, youth or beauty. It would be a world for people who, unable to attend to their surroundings, find themselves pushed and pulled by their cravings: good consumers, perhaps, but restless people lost in a world of merely functional things, caught up in a *samsara* of desire and plastic.

Much more could be said here about the dangers of consumerism. But for present purposes it is enough to note the relation between our material circumstances and our capacity to develop and exercise attention. In short, the relation is, I suggest, this. Merely functional things do not hold the attention; in fact many of them have been specifically designed not to hold the attention. And so in a world dominated by mass-produced functional things, attention would be hard to exercise and therefore hard to develop. The virtue could be more easily developed in a world of things that display their particularity and integrity and thus hold the attention, a world of what Heidegger, following Rilke, calls *things*. So, for instance, it could be developed and exercised in the presence of art, for artworks – or at least, good ones – have a singular capacity to invite and sustain attention. I take it that this is Heidegger's point when in 'The Origin of the Work of Art' he maintains that although a great work of art, like Van Gogh's painting of a peasant's shoes, can intimate both the world of the peasant woman and the self-secluding 'earthy' base on which it is established, all this is only evident in the *painting* of the shoes. The peasant woman, by contrast, 'simply wears them' (*BW*: 160). More generally, he adds, the more a thing discloses itself as 'handy', the more inconspicuous its 'being' becomes – the less able it is to hold the attention (cf. *BW*: 190).

If artworks invite attention, then the virtue can also be developed and exercised in the presence of nature.[4] It is true that many natural things are, in a certain sense of the phrase, mass-produced. One acorn, for instance, seems to be pretty much identical to any other. Nonetheless, it is often hard to see natural things as *merely* functional. To be sure, one might find some use for a piece of driftwood or a specimen of bracket fungus, or an acorn, for that matter; indeed one might come to regard these things as mere resources (and I will consider this possibility below). However, in contrast to television sets, plastic cups and other mass-produced functional items, it tends to be difficult to regard them in this way. For unlike television sets and the like, natural things have not been designed to fulfil any human purpose and so there typically seems to be more to them than can be comprehended in instrumentalist or functionalist terms. They more readily present themselves to us as things in their own right, as having what Nyanaponika calls an 'evanescent individuality' of their own.[5] And for this reason, I would suggest that, just as on Murdoch's account great art 'invites unpossessive contemplation' (2007: 83), so natural things can invite attention. In this way, the presence of nature can help to reinvigorate what Annie Dillard calls our capacity to see things 'truly' – and, it should be added, our capacity to listen, to touch, to smell and to taste (see further, Dillard 1998: 120–1). Indeed for Murdoch this is one reason to cherish the 'sheer alien pointless independent existence of animals, birds, stones and trees' (2007: 83).[6]

4.3 Responses to some objections

To recap, it is a good thing, morally speaking, to let beings be, and a settled disposition to do this counts as a moral virtue (attention). When it is done well, phenomenological reflection necessarily involves attention. In this sense, it is a good thing to do phenomenology (or at least a good thing to do it well). Mass-produced things tend to militate against the development and exercise of attention. Accordingly, the virtue is best developed in other contexts – in the presence of art objects, for instance, or (the possibility we have been considering) in one's dealings with natural things. In short, then, I have suggested that in societies dominated by the forces of mass production (but not, perhaps, in all cultures), natural environments have a distinct value: they disclose themselves as places in which we might develop and exercise the virtue of attention. In fact, as the forces of mass production come increasingly to dominate our world, such environments will become more and more valuable on account of their capacity to invite and so to foster attention. This provides us with one reason for conserving nature (though there will of course be others).[7]

Yet this increase in nature's value must be set against another, opposing tendency. For we find ourselves at a point in history where technology, the 'way of revealing' of which mass-production is one concrete manifestation, has even begun to shape our relations with natural environments. As the world becomes increasingly technological, in Heidegger's sense, more and more of it discloses itself to us as standing reserve. And as we saw in Chapter 3, this is even evident in our relations with the natural world. Nature is increasingly revealing itself to us as a sink of mere resources – not only as timber and cash crops, but as a provider of natural services and a store of natural capital. And as this trend continues, the capacity of natural things to invite attention will wane. Indeed, if at some point in the future we were to find that nature had come entirely to disclose itself as merely functional, as nothing but standing reserve, then it would have ceased to invite attention, and its value, as described above, would have diminished to zero. Hence at this point in history we find ourselves at the intersection of two opposing trends. As the value of nature increases, so we find ourselves less and less able to see nature for what it is. As natural places become more and more valuable as contexts within which we might be able to develop attention, so we find ourselves less and less able to see nature as anything more than a sink of resources. And to the extent that nature discloses itself to us in this way, it fails to invite attention, and so its value wanes.

Φ

I have argued that we ought to conserve natural things and natural environments because, to the extent that they disclose themselves as non-functional, they can help us develop and exercise the virtue of attention. But to some this appeal to the importance of attending to nature might seem problematic, even downright objectionable. For it might seem to represent a bourgeois approach to environmental ethics, an expression of the distinctively middle and upper class notion that the natural world is a spectacle for one's spiritual edification. In response, one could challenge the empirical claim that a love of nature is a bourgeois phenomenon. However, even if that claim proves to be correct, it would be wrong (an instance of the genetic fallacy) to suppose that the view must be flawed simply because it tends to be entertained by people of a certain social class. At the very least, that claim would have to be supported by independent arguments (of a Marxist variety, for instance).[8]

Φ

But the case set out above might seem objectionable for another reason. For it would seem to imply that we have a moral obligation to conserve natural environments because they can help us in our morally praiseworthy efforts to cultivate the virtue of attention. But to affirm this, is, it would seem, to regard nature as *instrumentally* valuable, as a *resource* for one's moral edification. And this proposition might seem troublingly anthropocentric.

Before responding to this objection, let me make it clear, right away, that I *do* believe that it is in our (i.e., we modern-day consumers') interests to conserve nature because natural environments tend to foster the development of attention. But let me add that, in keeping with the moral pluralism espoused in the previous chapter, I do not believe that this is the *only* reason we have for conserving nature. Only that it is *a* reason.

This granted, how is one to respond to the charge of anthropocentrism? The first thing to note is the familiarity of its form. Environmental philosophers, especially those towards the dark green pole of the environmental spectrum, tend to greet appeals to human interests with suspicion. Such appeals are regarded as being anthropocentric, expressions, perhaps, of a 'shallow' as opposed to 'deep' approach to environmental issues, or, still worse, of the hubristic tendency to regard nature as a sink of resources to be exploited for material gain.

Yet such blanket condemnations of appeals to human interests fail to convince, for an examination of what it means to have an interest reveals that interests come in many different varieties, some of which ought not to be condemned as anthropocentric.

This point is best conveyed by means of an example. So consider, first the case of a farmer trying to decide whether to fell a stand of trees growing on his land. It may be in the farmer's interests to fell the trees so that he might profit from selling the timber, but if such instrumentalist concerns move him to action, then he may be guilty of adopting a perniciously human-centred standpoint in his practical deliberations. He may be guilty of anthropocentrism.

But now let us suppose that another man, a local postman, finds the trees beautiful, so beautiful that he will often go out of his way to walk by them on his way home from work. The postman has an interest, an aesthetic interest, in the trees. Now some writers would maintain that, morally speaking, the postman's response is similar to that of the farmer. True, both men value the trees, but only because they both hope to benefit from them in some way. The farmer wants cash, the postman, a pleasurable aesthetic experience, but they both hope, as it were, to 'get something out' of their engagement with the trees. Although we might be tempted to condemn one man and applaud the other, the interests of both men are essentially alike.[9]

But this rather jaded interpretation of aesthetic appreciation cannot be upheld. For to suppose that aesthetic appreciation is inherently anthropocentric (and thus that the farmer's response is morally equivalent to that of the postman) is to conflate two meanings of 'anthropocentrism'. In one sense, aesthetic judgements are inherently anthropocentric, for if something is judged to be beautiful or ugly, that judgement must presuppose some relation to human interests.[10] But this is anthropocentrism of a morally unobjectionable sort. By contrast, the anthropocentrism of which environmental thinkers should be wary involves, not the merely formal claim that a particular judgement has meaning only because it connects in some way with human interests, but the substantive claim that something has value only to the extent that it satisfies a particular subset of human interests, typically interests in exploiting the non-human world for material gain. Aesthetic judgements such as the postman's may be anthropocentric in the first sense, but not in the second, morally problematic sense.

The charge of anthropocentrism fails, therefore, when it is directed at the aesthetic appreciation of nature. Interests in natural beauty, for instance, are not inherently anthropocentric, and ought not to be dismissed as such. More generally, it would be a mistake to dismiss all appeals to human interests as anthropocentric. It may be in our interests to relate to nature in a certain way, but it remains a further question whether we ought to be condemned as anthropocentric for doing so.

With this in mind let me introduce a third figure, a 'green mystic' who values the farmer's stand of trees, neither because he hopes to profit from them nor because he finds them beautiful, but rather because he finds that, under the trees' branches, his capacity for attention is intensified.

In judging whether this man ought to be condemned as anthropocentric, we must consider whether his interest in developing and exercising the virtue of attention counts as objectionable (like the farmer's) or unobjectionable (like the postman's). At first sight, it might seem that the second interpretation is correct. For, unlike the farmer, the mystic is presumably concerned, not with what nature can do *for* him, but with what it can do *to* him (cf. Muir 2007: xxiii). He does not seek to use nature, but to be transformed by it, and there might not seem to be anything anthropocentric about this.

But the mystic's desire to be transformed is not enough, in itself, to exonerate him. For the key factor is not whether he seeks to use nature or to be transformed by it, but his *motivation* for seeking out natural places. The interests of the man who goes to the woodland in the hope of being transformed must be distinguished from those of the man who goes to the woodland for other reasons but just so happens to be transformed by the encounter. Our moral assessment of the first set of interests will reflect our moral assessment of the transformation sought. If the man hopes to be transformed in some morally edifying way, then his desire to be so transformed ought not to be censured. If he does not seek to be transformed in some morally edifying way, his interests will be deemed morally unproblematic at best, or morally objectionable at worst. But the interests of the man who just so happens to be transformed by his walk in the woods must be judged differently. It may be in such a man's interests to go to the woods, even if his forays under the trees are not motivated by a conscious interest in being edified. He might be edified by his encounters with nature even if he does not seek out such encounters in the hope of being edified. Indeed, when it comes to attention, it seems likely that a desire to develop and exercise the virtue will prove counterproductive. For to attend to a thing, in the special sense described above, is to seek to do justice to the thing itself, and to do this, one must divest oneself of self-centred concerns. Thus Murdoch insists that the 'loving gaze' requires an 'unselfing' on the part of the subject (2007: 82), while Heidegger maintains that in order to let beings be, one must regain the 'essential poverty of the shepherd' (*BW*: 245).[11] The upshot of this is that in order genuinely to attend to a thing, one must relinquish any desire to be edified by attending to it. In this way, a desire to develop or exercise attention might *militate against* the development or exercise of that virtue (see further, James 2006). Hence it is entirely possible that our 'green mystic' is motivated neither by self-interest nor by some anthropocentric interest in exploiting the non-human world for the benefit of us humans.

For these reasons, it is not anthropocentric to claim that natural things and natural places have therapeutic value because they can help us develop attention. But there is another reason why the charge of anthropocentrism fails to convince. For it presupposes the false notion that if attention is a virtue, then the only reason it can be a virtue is because it contributes in some way to human well-being. It assumes, in other words, that if attention is a virtue then it must be a virtue for eudaimonistic reasons. But this is not what I have argued. While I have suggested that attention is good for us in the sense that it is an integral component of human well-being, I have also suggested that it is a virtue for what might be described as other-regarding reasons as well, for reasons, that is, that may be specified without reference to the well-being of the attentive agent. I have suggested that it is a good (i.e., virtuous) thing to exercise attention because doing so allows beings to disclose themselves as they truly are, because, in Heidegger's apt phrase, it lets them be. And to let nature be is to be focused neither on one's own well-being nor on the well-being of humans as opposed to non-humans. It is simply to attend to whatever part of nature is at hand.

Φ

I have argued that my attention-based approach to environmental ethics ought not to be condemned as either bourgeois or anthropocentric. Yet these are not the only objections that could be raised against it. For instance, some more practically minded critics might worry that reflecting on the merits of attending to nature might distract us from the pressing issue of responding to the myriad environmental challenges facing us today. After all, something needs to be done in response to environmental issues, and fast! To reflect on the virtue of attention – one worries that that might be to fiddle while the world burns.[12]

There is something to this criticism, of course. We (or rather, primarily, people in some of the world's poorest countries) are facing an environmental crisis. At this very moment, ice is melting and seas are creeping inland, rainforests are shrinking and cities are sprawling, species are being irretrievably lost. But there is another crisis too. As I write, *things*, in Rilke's sense or Heidegger's, are disappearing. When we look around us we have to look hard, and harder each year, to find particularity and integrity. Instead, we seem increasingly to be faced with a world of *Bestand*. More and more, where we once saw things, we see only ourselves. Things are fading, dissolving into a shifting miasma of human cravings.

True, some things persist: cherished things, things with which we have lived, artworks, and of course natural things. But with respect to the last of these there are two dangers. The first is that natural things might literally disappear, that over time green and growing things will be killed off and tarmacked over, replaced by concrete, plastic and metal. But the second danger is that the natural things and natural places that remain will come to disclose themselves in merely functional terms, as natural capital, perhaps, or as the providers of natural services. The second danger, then, is that natural things will come to reveal themselves to us as mere resources.

Our task, then, is to conserve nature in two ways. Through action we must ensure that natural things and natural places remain, that the world does not become, in material terms, one of concrete, plastic and metal. But we must also ensure that nature continues to disclose itself to us *as* nature, and not merely in functional terms. We must let it be, not merely for our sake or for the sake of our well-being, but for the sake of nature as well. And this is the work of attention.

Notes

1. One way of putting this would be to say that Heidegger's releasement amounts to a kind of 'moral vision', in Murdoch's sense. However, visual metaphors may not be the most appropriate in this context. After all, while releasement takes the form of a kind of receptivity to the world, vision tends to be construed as an activity of a subject. If a woman sees the truth, that, we think, is her accomplishment. In seeing the truth, she acted, and she deserves the credit. (One thinks here of some Medieval theories of vision, according to which it was the result of rays emitted by the eye.) In the light of these observations, one might conclude that the meaning of releasement is better conveyed by means of an analogy with *listening* (see Tuan, quoted in Saito 2004: 141; also Levin 1989).

2. The Buddhist view on these matters is close to that of Murdoch (which is unsurprising, given Murdoch's admiration for the religion). For Buddhists, to be mindful is to be acutely aware of a variety of phenomena, from thoughts to bodily sensations (the rise and fall of the breath, for instance). In the context of Buddhist practice, a dispassionate awareness of these phenomena is thought to foster a sense of their transience and, accordingly, freedom from attachment (see further, Gowans 2003: 189–91). But as ever in Buddhism, the ability to do this is not regarded as being of benefit only to the practitioner. Mindfulness is thought to go hand in hand with a caring and attentive attitude towards others and, indeed, towards the world as a whole. So, as with Murdoch, attending to reality is thought to be, in part, a moral achievement, and the virtues thus brought into play are regarded as moral virtues (see further, Cooper and James 2005: 101–3).

3. It is true that, seen with the right kind of eyes, the television set could disclose itself as being an object in its own right. So, for instance, a devotee of the *objet trouvé* school of art could see it as an art object. However, the fact remains that mass-produced things like the television set tend to inhibit this way of seeing (which is perhaps one reason why works such as Duchamp's *Bottle Rack* were considered so radical).

4. Heidegger himself came to recognise that things other than artworks can invite attention. So in later essays such as 'The Thing' (*PLT*: 165–86) and 'Building Dwelling Thinking' (*BW*: 143–212), he proposes that

commonplace 'things', such as earthenware jugs and even 'heron and roe, deer, horse and bull' (*PLT*: 182) can serve to 'gather' a world and so reveal the mysterious workings of Being, on which all things depend.

5. I do not mean to suggest that art objects or natural things must be entirely non-functional. The former can certainly be functional. (A Japanese tea bowl, for instance, might be at once useful and a thing of beauty.) And natural things, for their part, can disclose themselves as both functional and non-functional. (Recall Wordsworth's Michael. For him, the skies and the soil are ready-to-hand, yet they do not disclose themselves as mere resources. They are what he lives with, not merely what he uses.) Nonetheless, whereas mass-produced items often disclose themselves as merely (or almost merely) functional, art objects and natural things do not. Something that has come to disclose itself as *merely* functional cannot be an art object. And in a certain sense, it cannot qualify as 'natural' either (a point I return to towards the end of this chapter).

6. Two caveats. First, what is important here is not exactly that a thing is natural (in the sense explained in the Introduction), but that it is *perceived* to be so. However, it is reasonable to assume that our judgements of what is natural usually coincide with what really is natural. Second, while I have suggested that all natural phenomena will have value to the extent that they invite attention, this value may be overridden by other considerations. So, for example, if I am lost on a mountainside, my sense of the mountain's indifference to my concerns, which in other circumstances could prove therapeutic, will be overridden by other factors, notably my desire to survive.

7. In view of the pluralism defended in Chapter 3, it must be added that even if nature has value in this sense, it will be morally significant for other reasons too (on account of the bonds we have with it, the meaning it embodies and so forth).

8. Cf. Eliot Sober's claim that although environmental concerns 'become important to us only after certain fundamental needs have been satisfied', they are not for this reason 'frivolous' (2002: 156).

9. Thus some writers have argued that an aesthetic approach to nature 'confirm[s] our anthropocentrism by suggesting that nature exists to please as well as to serve us' (R. Rees, quoted in Carlson 1998: 125).

10. Admittedly, one can conceive of non-human beings capable of aesthetic appreciation. But in the absence of evidence for non-human connoisseurs, it is reasonable to assume that aesthetic judgement is inherently anthropocentric in the sense explained.

11. 'Poverty' here refers to the spiritual poverty realised by mystics, the state of the man who 'wants nothing, knows nothing, and has nothing' (Meister Eckhart, quoted in Merton 1993: 222).

12. For a forceful statement of this kind of view, see Frodeman 2006.

5
Beyond the Human

5.1 Realism and anthropocentrism

The preceding four chapters have set out a phenomenological approach to several key issues in environmental philosophy. Yet the very notion of developing a phenomenological approach to such issues might seem wrong-headed. For it might seem that phenomenological approaches are inherently human-centred or anthropocentric and hence at odds with the concerns of environmental thinkers.

This anthropocentrism might seem to be evident in the phenomenologist's commitment to a certain variety of metaphysical anti-realism. As we saw in Chapter 1, phenomenologists maintain that the world is internally related to us, which is to say that whatever 'shows up' for us as a being does so in the light of certain features of us – our interests, attitudes, perspectives, practical concerns and so forth. It is for this reason that Heidegger writes that '[i]f no Dasein exists, no world is "there" either' (*BT*: 417). Likewise, for Merleau-Ponty the world is one in which 'every object displays the human face it acquires in a human gaze' (*WP*: 70). The claim here is not simply that the world as *experienced* reflects human concerns (even the most committed realist would be happy to admit that). It is that something that bore absolutely no relation to our lived experience would, of necessity, elude our comprehension. This, I take it, is what Merleau-Ponty means when he claims that '[t]he perceived world is the always presupposed foundation of all rationality, all value and all existence' (*PrP*: 13). And it also seems to be what he is suggesting in the following passage:

All my knowledge of the world, even my scientific knowledge, is gained from ... some experience of the world without which the symbols of science would be meaningless. The whole universe of science is built upon the world as directly experienced

(*PP*: viii).

The suggestion here is (partly, at least) that scientific theories are meaningful if and only if they bear in some way upon our lived experience of the world. And this is thought to hold true even of widely accepted theories such as the theory of evolution by natural selection. The point here is not that this theory, for instance, is *merely* a theory (as a creationist might allege), but that it only makes sense to us because of the fact that it relates to some more basic encounter with the world. (It is a further question whether it is a *good* theory, which is to say that it remains to be shown whether neo-Darwinian selection provides a better explanation of natural phenomena than theories that appeal to genetic drift, say, or developmental constraints.) The important point for our purposes is that, if Merleau-Ponty is right about this, then talk of how the world is in itself, independently of human interests, attitudes, practical concerns and so on is without sense:

To our assertion ... that there is no world without an Existence [a human existence] that sustains its structure, it might have been retorted that the world nevertheless precedes man. [But] what precisely is meant by saying that the world existed beyond any human consciousness? An example of what is meant is that the world originally issued from a primitive nebula from which the combination of conditions necessary to life was absent. But every one of these words, like every equation in physics, presupposes *our* pre-scientific experience of the world, and this reference to the world in which we *live* goes to make up the proposition's valid meaning. Nothing will ever bring home to my comprehension what a nebula that no one sees could possibly be.

(*PP*: 432)

'Poor Maurice!', one can imagine the realist replying. 'Not to be able to conceive a nonhuman world! His life must be poorer as a result'. But Merleau-Ponty is not denying that he can comprehend the nebula. Of course he can – he is writing about it, after all. He is suggesting that our talk about even so ostensibly non-human a thing as the nebula fails to refer to how the world is in itself, for insofar as such talk has meaning it must connect with our lived experience in some way.[1] It is in this sense that the phenomenologist's position is anti-realist.

Φ

The theory of meaning implied by the 'nebula' passage can be challenged (see, for instance, Thomas Baldwin's objections in *BWr*: 20). However, it is not my aim here to defend the phenomenologist's views on this specific issue, merely to note that those views appear, on the face of it,[2] to be anti-realist.

This preliminary conclusion would no doubt perturb many environmental thinkers. After all, to reject the idea that there is a way the world is in itself, independent of human concerns, would seem to make *Homo sapiens* the measure of all things, and that Protagorean thesis would seem to exemplify the kind of anthropocentric hubris environmental thinkers have traditionally opposed. This is why environmental thinkers like Holmes Rolston object to anti-realist claims to the effect that it is '*We* [i.e., us humans] who cut up the world into objects when we introduce one or another scheme of description' (1997: 53).[3] Surely, Rolston contends, 'we do not think that lion-objects come into being when we humans arrive and cut up the world into such objects' (1997: 54). On the contrary, '[t]he Earth-world was quite made up with objects in it long before we humans arrived' (1997: 55). For similar reasons, other environmental thinkers take issue with the anti-realist tenor of some postmodern writings. Thus Peter Coates suggests (rather implausibly) that for many 'thoughtful Greens' postmodernism represents the 'greatest threat to nature today' (1998: 184) – a view echoed by another 'environmental realist', Paul Shepard:

> There is an armchair or coffeehouse smell about [postmodernism]. Lyotard and his fellows have about them no glimmer of the earth, of leaves or soil. They seem to live entirely in a made rather than a grown world; to think that 'making' language is the same as making plastic trees, to be always on the edge of supposing that the words are more real than the things they stand for.
>
> (1995: 20)

In Shepard's view, postmodernism is 'more like a capstone to an old [anthropocentric] story than a revolutionary perspective' (1995: 24–5). What is needed, to quote another environmental realist, is not another postmodern deconstruction of nature, nor, more generally, another dose of anti-realism, but 'a real appreciation of a genuine Otherness, a world not limited by what we make of it' (Clark 1994: 125).

5.2.   The alterity of nature

The environmental realist's charge is that anti-realism cannot do justice to the sense in which nature exists independently of our concerns, and that, for this reason, a commitment to anti-realism is at odds with a proper moral concern for the natural world. And since the theses of Heidegger, Merleau-Ponty et al. are regarded as being anti-realist, all this is thought to militate against phenomenology as well.

The aim of this chapter is to determine whether these criticisms are justified, to determine, in other words, whether a phenomenological approach can do justice to the independent reality of the natural world or whether it is, by contrast, perniciously anthropocentric. I will contend that while, in one sense, phenomenology is anthropocentric, this human-centeredness is not of the pernicious sort that ought to perturb environmental thinkers. And in addition to this, I will argue that in some of its forms phenomenology is not anthropocentric at all.

Φ

One thing must be noted right away: to deny that sense can be made of the notion that there is a way the world 'anyway is', independent of human concerns, is not necessarily to suppose that the world is merely a *product* of our understanding. Rolston maintains that 'perception is only intelligible if it is contact with objects and events out there', that is, in the world (1997: 56). The phenomenologist would agree. Thus Merleau-Ponty maintains that 'I have no doubt that I am in communication with [the world]' (*PP*: xvii). The world is not merely a product of human understanding; on the contrary, it has an otherness or alterity to it, a side that could not have been constituted by us (cf. Madison 1981: 31).

On the phenomenological account, then there is a sense in which the world exists independently of us. Indeed, this independence is thought sometimes to disclose itself in experience. For instance, in his 1935 lecture, 'The Origin of the Work of Art', Heidegger claims that a great work of art reveals not only a world, but also a concealed, or rather conceal*ing* dimension, which he calls the 'earth'. 'World' here has essentially the same meaning as it did in *Being and Time* – it continues to stand for a field of significance grounded in the concerns of Dasein. Moreover, Heidegger still accords a primacy to the equipmental aspects of particular things: in his famous discussion of Van Gogh's painting of a peasant's shoes, for example, he refers to the shoes as *Zeug*, equipment. However, by the time of 'Origin' Heidegger had come to think that 'this equipment belongs to the *earth*' (*BW*: 159).

Heidegger tries to illuminate the relation between world and earth by discussing the idea of 'reliability'. He claims that the reliability of the peasant woman's equipment points to the hidden connection between the tools with which she is unreflectively familiar – the shoes, the hoe, the gloves and so on – and the 'earthy' foundation on which her work is founded. As reliable, he suggests, the equipment must be intimate with the earth: 'by virtue of this reliability the peasant woman is made privy to the silent call of the earth' (*BW*: 160). Thus he writes that the shoes intimate a world transcending the human: the 'furrows of the field swept by the raw wind', the 'dampness and richness of the soil' and the 'loneliness of the field path as evening falls' (*BW*: 159). The earth, here, seems to stand for a dimension conspicuously absent from *Being and Time*, one that encompasses all that *conceals* itself from the concernful dealings of Dasein.

Merleau-Ponty also discusses the ways in which things disclose themselves as existing independently of us. He maintains that, when it is viewed in a particular way, a thing can disclose itself as 'a thing in itself' (or more precisely, a 'thing-in-itself-for-us'), as 'aloof from us and ... self-sufficient', as 'hostile and alien ... a resolutely silent Other' (*PP*: 322). This, at least in part,[4] is to say that it exceeds our understanding: '[w]e find that perceived things, unlike geometrical objects, are not bounded entities whose laws of construction we possess *a priori*, but that they are open, inexhaustible systems which we recognize through a certain style of development, although we are never able, in principle, to explore them entirely, and even though they never give us more than profiles and perspectival views of themselves' (*BWr*: 5–6). So reflecting on the wallet before me, I find that this very disclosure includes, as an integral part of it, the possibility of further disclosures. I see only one side of the wallet, yet this perception includes, as a 'horizon', a sense of its other, hidden face (cf. *PP*: 68). And these horizons are multiplied through a kind of entirely natural, though seldom noted, synaesthesia. I am looking at the wallet, yet I perceive how it would feel were I to touch it. And I am aware, too, of the smell of the leather, even its rich, salty taste. The wallet, in short, is open to unending exploration; it has, as Merleau-Ponty puts it, an 'inexhaustible depth' (*VI*: 143). And one might think that this is a result with which the realist should be happy. The thing is not a product of our understanding. On the contrary, it outstrips our consciousness of it; it is what Merleau-Ponty calls a 'transcendent' thing (e.g., *PP*: 369).

Φ

I do not mean to suggest that Heidegger's 'earth' is equivalent to Merleau-Ponty's 'thing-in-itself-for-us'. On the contrary, the two notions are in certain respects quite different. What I would like to note is that although for both thinkers the world is a 'human' one in the sense that things disclose themselves in the light of our concerns, it is also one in which things can reveal themselves as having a non-human aspect, a reality independent of that which is revealed to us humans.

Let us grant for the sake of argument that these general claims can be justified. Can they be used as the basis for an account of the independent reality of the *natural* world?

The fact that Heidegger calls the non-human dimension of things *earth* (*Erde*) is surely significant, as is his decision to make his case by referring to a peculiarly rustic context of ploughed furrows and field paths. Likewise, Merleau-Ponty maintains that the ungraspable depth of the thing intimates, as a horizon, the presence of a single, wider world – 'the *natural* world' – from which all things emerge and into which they recede: 'one being, and one only, a vast individual from which my own experiences are taken, and which persists on the horizon of my life as the distant roar of a great city provides the background to everything we do in it' (*PP*: 328).

But one must not read too much into these observations. For one thing, it may be noted that Heidegger chooses to illustrate what the earth is by referring to a selection of *artefacts*, the peasant woman's various tools. Similarly, there is no good reason to equate Merleau-Ponty's 'natural world' with that of the environmental thinker. On the contrary, according to the account set out in *Phenomenology of Perception* the 'one being' intimated in one's perception of a particular thing refers to more than just non-artefactual things. The natural world, in this sense, remains a horizon throughout my life, whether I am hacking my way through the jungle or wandering through an air-conditioned shopping mall. The environmental thinker, of course, sees things differently, being typically concerned with the natural-as-opposed-to-the-human world.

Nonetheless, it is surely easier to appreciate the independent reality of things, their 'earthy' or 'nonhuman' aspect, in some situations than in others. So while Heidegger chooses to convey the meaning of earth by referring to certain artefacts, it is significant that the artefacts to which he refers are *not* mass-produced. Indeed in several of his later essays he claims that earth is evident in *things*, by which he tends to mean what would usually be referred to as a certain *kind* of thing, namely one that, through years of use, has come to occupy a key role in the life of a particular community. So while, for Heidegger, an earthenware jug could be a thing, a Tupperware one probably could not.

Merleau-Ponty, for his part, maintains that beings will be less likely to offer up their non-human faces when they disclose themselves in the context of our everyday practices: 'Ordinarily we do not notice [their alterity] because our perception, in the context of our everyday concerns, alights on things sufficiently attentively to discover in them their familiar presence, but not sufficiently so to disclose the non-human element which lies hidden in them' (*PP*: 322). The otherness of the thing only becomes evident 'if we suspend our ordinary preoccupations and pay a metaphysical and disinterested attention to it' (*PP*: 322; cf. *BT*: 102–3). There is no good reason to conclude on this basis that the equipment of everyday life will never reveal its otherness – perhaps a suitably attentive frame of mind would be able to discern the non-human element in anything (see further, Vogel 2003; Abram 1996: 64). Yet it seems likely that some things will be, as it were, less willing than others to yield up their alterity. Consider mass-produced artefacts, for instance. As we saw in the previous chapter, there is scarcely anything about them that points beyond the compass of human concerns. They tend to disclose themselves as what Keekok Lee has called 'the material embodiment[s] of human intentionality' (1999: 2). As David Abram puts it, they 'only reflect us back to ourselves' (1996: 22).

The non-human element to which Merleau-Ponty refers might therefore be particularly evident in things that are not merely 'functional'. Perhaps, as Heidegger suggests, the non-human element of artworks is especially perspicuous. And even more so, perhaps, that of natural things. After all, it has long been recognized that natural phenomena are on the whole better able to invite disinterested appreciation than artworks precisely because they disclose themselves as being less tied up with distinctively human concerns, about the character of the artist, say, or the intended message of the work (see further, Brady 2003: 128–9). Likewise, as we saw in Chapter 4, natural things might be better able to invite the disinterested attention needed on the part of the perceiver to discern their non-human aspect. A man intent on discovering the 'non-human element' in things, of which Merleau-Ponty writes, would be well advised to take to the woods, for that aspect of the world might be harder to overlook in such a context.[5] Or as Merleau-Ponty himself puts it in 'Cézanne's Doubt':

> We live in the midst of man-made objects, among tools, in houses, streets, cities, and most of the time we see them only through the human actions which put them to use. We become used to thinking that all of this exists necessarily and unshakeably. Cézanne's painting suspends these habits of thought and reveals the basis of inhuman nature [*le fond de nature inhumaine*] upon which man has installed himself.
>
> (*SNS*: 16; cf. *PP*: 324.)[6]

Φ

The phenomenologist therefore rejects the claim that the world is a product of our understanding. Indeed, she emphasises that the independent reality of things is sometimes evident in experience, when, for instance, one finds oneself privy to their 'earthy' or non-human dimensions. And to this extent the phenomenologist is able to acknowledge the independent reality or alterity of *natural* things. In fact, she is able to show why the independent reality of things may be *especially* evident in our dealings with nature.

Yet a realist would, no doubt, remain unconvinced. For one thing, the claim that the independent reality of things is most evident in our encounters with nature might strike him as banal (aren't natural things defined as such precisely because they seem to transcend human concerns?). For another – and more importantly – the realist might remain unconvinced by the phenomenologist's claim that her approach is *not* anthropocentric. After all, while Merleau-Ponty (for instance) rejects the anthropocentric notion that the thing is merely a product of human understanding, it remains the case that in *Phenomenology of Perception*, at least, this 'non-human', unconstituted part of the thing is only considered insofar as it shows up as a horizon in the perception of a human subject. It would appear that Merleau-Ponty is not interested in the more-than-human world *per se*; his concern is to elucidate the role that world might play in (human) perception. So although he writes of the thing as an 'in-itself-for-us', he is quick to add that it 'is inseparable from a person perceiving it, and can never be *actually* in itself because its articulations are those of our very existence' (*PP*: 320; my emphasis). However independent it may appear, the thing is in actuality the perspicuous embodiment of a secret 'communion' between perceiver and perceived (*PP*: 320). And the same holds true of the natural world in which it inheres: this, too, takes the form of a horizon of (human) perception (cf. *PP*: 328; cf. Barbaras 2001: 27).

Φ

So it would seem that although phenomenological reflection can uncover the non-human element in things, their alterity, this alterity will not be *radically* non-human since it will necessarily disclose itself in relation to our concerns. The thing might very well disclose itself as 'hostile and alien', as 'a resolutely silent Other' and so forth, but that signifies what, to adapt Merleau-Ponty, one might call a non-humanness-for-us, and not a non-humanness of a radical sort (cf. *PP*: 322).

For the realist, of course, this is all highly unsatisfactory. The notion of a 'non-humanness-for-us' would seem to be altogether too human-centred to be acceptable. What the realist wants, what he thinks we inveterate anthropocentrists *need*, is to acknowledge that there is a way the world 'anyway is', independent of human perspectives, attitudes, practical concerns and so forth (see further, Cooper 2002: 1*ff.*). And what the *environmental* realist wants, for his part, is for us to acknowledge that there is a way nature 'anyway is', a way it is 'in itself'.

But these demands are unreasonable. If Merleau-Ponty and other phenomenologists are correct – and on this point I think they are – the notion that there is a way the world 'anyway is' is without sense. For in order to count as a world at all, any world must connect in some way to our concerns, broadly construed. Of course realists such as Holmes Rolston are right to say that the world was 'quite made up with objects in it long before we humans arrived' (1997: 55). But the world to which they refer is not *radically* non-human since it embodies the conceptual schemes, practical concerns and so forth that are brought into play in its making sense to us. In this respect, even the pre-human world is indelibly 'human'. And to describe that world is not to describe a way the world 'anyway is', for such a radically objective state of affairs could not count as a world at all, still less one that we could grasp in thought or imagination.[7]

We can therefore scratch off one reason for being suspicious of the phenomenologist's account. True, that account cannot furnish us with an account of how nature is in itself, but since any such notion is incoherent this result is no cause for regret. If the phenomenological account is anthropocentric, then this anthropocentrism is of a merely formal and entirely necessary sort.

And there is another problem with the environmental realist's reasoning. If the phenomenologist is correct, there is no sense in which the world is in itself, independently of our concerns. Our 'outlook' on the world is, in this sense, indelibly human. But it is a further question whether that outlook ought to be regarded as *distinctively* human in all respects. After all, a man's outlook on the world is in an entirely trivial sense that of a man. But it is a further question whether it ought to be regarded as *distinctively* male in all respects. Indeed, it is a further question that must be answered in the negative, for much that can be said of a man's outlook on the world also holds true of the outlooks of people in general, male or female, young or old. Similarly, to say that we humans necessarily experience a world traced out in terms of our attitudes, perspectives, practical concerns and the like is not to say that every element of our experience reflects *distinctively* human concerns. On the contrary, as we saw in Chapter 2, there are good reasons to think that overlaps exist between our outlook on the world and the outlooks of certain non-human animals. One could say that although we necessarily see the world from our perspective, that perspective may be one that we share, to some extent, with certain non-human others.

This result has some interesting implications for our attempts to understand the independent reality of nature. It is sometimes contended, against a certain variety of metaphysical realism, that reality as it is 'in itself' must be in principle unknowable, like Kant's 'thing-in-itself'. And a similar charge could be levelled at the environmental realist's claim to have had some insight into what nature is like 'in itself'. Now arguments of this sort may or may not work, but if they are found to have force they will only militate against attempts to *represent* nature in itself. They will only debar us from knowing what nature is like 'from the outside', as Schopenhauer might have said. The possibility of a 'way from within', of a direct and non-representational insight into nature, remains open.[8]

This requires explanation. I have suggested that our outlook on the world overlaps that of certain non-human animals, or, in less subjectivist terms, that to some extent humans and certain non-human animals have a common way of inhabiting the world. Now if such overlap exists, then interrogating our own being-in-the-world could afford us an insight into that part of nature that manifests itself, not only in our way of being, but in that of at least some non-human animals as well. As Scott Churchill notes, this would not be merely theoretical 'knowledge about' nature, but 'knowledge by acquaintance' (2008: 175). It would be to appreciate the workings of a nature that moves not only in us but in non-human beings too.

These are bold claims, and they deserve more argument than I can give them here. But regardless of whether they are found to hold water, they clearly ought not to be dismissed as anthropocentric. Whatever else they are, they are evidently not perniciously human-centred.

Φ

The phenomenological account should not, therefore, be dismissed as inherently anthropocentric. Indeed there are reasons for thinking that it may be particularly well suited to shedding light on the independent reality of the natural world.

Recall the debate with which we began. Environmental realism, we saw, tends to be regarded as a philosophical position, one, in other words, that is articulated and defended by academic philosophers. Yet this way of framing matters can be misleading. For although environmental philosophers (at least when they have their professional hats on) may be much exercised by the question of whether environmental realism is a tenable position, many other green thinkers will be left cold by such debates. For many such thinkers, the reality of the natural world is not merely a proposition to be articulated and defended, but something that is immediately evident, at least some of the time.

One of the great virtues of phenomenology is that it has the potential to do what more abstract epistemological and metaphysical debates on environmental realism cannot, namely, to deepen our understanding of such moments, to show, that is, what it means to perceive, in an immediate and visceral way, the independent reality of the natural world. And phenomenology can achieve this precisely because it is focused, not on abstract conceptions of an objective world, but on how the world discloses itself to beings like us. After all, even the 'mind-independent' world so beloved of realists could only matter to the extent that it bears upon our lives. Any part of reality that failed entirely to connect with our lived experience would, like Merleau-Ponty's nebula, be nothing to us, an idle wheel in our understanding.

Φ

The natural world discloses itself to us as an independent reality in many ways. In the following, I focus on two of the most interesting: first, the way in which the world can disclose itself as an object, not just for us, but for non-human subjects as well; and, second, the disclosure of nature as a realm indifferent to human concerns.

To introduce the first of these topics, I begin, not with nature, but with the writings of a thoroughly urbane and cultured phenomenologist, Jean-Paul Sartre. In *Being and Nothingness*, Sartre describes the experience of sitting in a park and then noticing another man passing by. In his characteristically melodramatic style, he explains how with the appearance of this man

an object has appeared which has stolen the world from me. ... Everything ... is fixed in the direction of the new object. The appearance of the Other in the world corresponds therefore to a fixed sliding of the whole universe, to a decentralization of the world which undermines the centralization which I am simultaneously effecting.

(*BN*: 255)

For Sartre, the man passing by is a constituent of the park-world, an object among other objects: he is shorter than that tree, closer than that bench, moving more slowly than that pigeon and so on. Yet he is also a subject, and when Sartre realises this, his entire world is transformed. The world is now not merely a private show for Sartre; it is for another subject too. The bench is an object of experience for someone else as well. The grass is green, not just for Sartre, but also for this man; indeed Sartre acknowledges that the particular shade of green which the man sees 'turns toward [him] a face which escapes me' (*BN*: 255).

Sartre interprets these observations in a rather downbeat fashion. The reader is left in no doubt that the appearance of the Other is, broadly speaking, a bad thing, one that results in one's world being 'stolen', and so forth. But this negative tone reflects nothing more than the bleak and antagonistic conception of interpersonal relations set out in *Being and Nothingness*, and it need not concern us here. More interesting for our purposes is the fact that the subject referred to in Sartre's example is *human*. This should come as no surprise. For Sartre, there are no animal subjects, and so all the subjects referred to in Sartre's works are human. But as we saw in Chapter 2, this sweeping rejection of animal consciousness cannot be justified. (As Kohák once wrote, 'only from a very great conceptual distance could one mistake a porcupine for a boulder and lump them both together under the common label of *l'être-en-soi*' (1984: 75).) And once this is conceded, it is but a short step to the conclusion that the 'decentralization' to which Sartre refers can be effected, not only by the presence of other humans, but by that of animals too. The tree is not just there for us humans, but also for the non-human inhabitants of the park. The park bench is an object of experience not just for members of the species *Homo sapiens*, but for a whole host of furred and feathered beings as well.

To appreciate the presence of non-human subjectivities is to have one's world transformed. For one thing, it affects one's perception of the world's *reality*. As we saw above, if a thing is perceived to be real it will disclose itself as having certain 'horizons'. I see only one side of the tree, yet this perception includes, as a horizon, a sense of its other, hidden side. But the possibility of other perspectives on the thing implies the possibility of other subjects. So, for instance, to see the tree as real is, among other things, to experience it as having sides which, though hidden from me, would be plainly visible to other observers standing in other positions relative to it. In this way, the tree discloses itself to me as exceeding my experience of it. It discloses itself as real (see further, *PP*: 338–9).

These remarks are entirely general. However, in the following passage David Abram explicitly connects the presence of non-human subjects with our sense of the reality of the *natural* world:

> I sense that that tree is much more than what I directly see of it, since it is also what the others whom I see perceive of it; I sense that as a perceivable presence it already existed before I came to look at it, and indeed that it will not dissipate when I turn away from it, since it remains an experience for others – not just for other persons, but … for other sentient organisms, for the birds that nest in its branches and for the insects that move along its bark. … It is this informing of my perceptions by the evident perceptions and sensations of other bodily entities that establishes, for me, the relative solidity and stability of the world.
>
> (1996: 39)

Abram is here referring not only to a world that contains non-human things, but to one that is *for* various non-human others, to a world which, for this very reason, is experienced by us as solid and stable, as real.

To appreciate this is to find oneself in a world that is not only more real but also stranger, more alien. In *Crow Country*, Mark Cocker describes how reflecting on the experience of rooks transformed his perception of the Yare valley in East Anglia. The landscape became 'at once smaller but freer, more unified and yet more fluid'. The air now resolved itself into a network of airborne causeways, 'tunnels of air' connecting areas of rookish significance: rookeries, roosts, worm-rich fields. And these areas now seemed to burn red, as if revealed through thermal imaging equipment. 'Suddenly', writes Cocker, 'the Yare valley had become a completely different landscape. It was not mine, it was not even *ours*. It was theirs' (2008: 132).

Φ

So natural things can disclose themselves as objects for both human and non-human subjects. Yet we ourselves can also show up as objects in the eyes of others. In *Being and Nothingness*, for instance, Sartre discusses the existential significance of being observed by others, suggesting that the 'relation' 'Being-seen-by-the-Other' is integral to one's being-in-the-world (*BN*: 257). There is no need to rehearse the details of his argument here. His conclusion, however, is that feelings such as shame and pride testify to our being-with others and thus militate against certain solipsistic anxieties, for in such feelings one 'live[s] ... the situation of being looked at' by the other (*BN*: 261).

Now Sartre, like Heidegger, does not countenance the possibility of a being-with animals, and this is reflected in his decision to focus on the feelings of shame and pride. Arguably, one can only feel ashamed or proud before another human (and perhaps God). But there is no good reason to suppose that the others by whom one can be seen, smelt, heard, tasted or touched must all be human or divine. For example, David Abram describes how, encountering a Himalayan lammergeier vulture, he felt his 'skin crawl and come alive, like a swarm of bees all in motion ... stripped naked by an alien gaze' (1996: 24). And one does not need to travel into the wilderness for such experiences. In the forest at night – even one that is small and familiar by day – the feeling of being watched can be palpable, sometimes overwhelming. Yet here the gaze one feels need not be human. One's being-with others may be conditioned by the primeval understanding between predator and prey (see further, Hatley 2004). Indeed, even the most familiar of creatures, our pets, can sometimes fix us in an unnerving, almost predatory gaze. (Recall the discussion of Derrida and his cat in Chapter 2.)

Φ

So the world – the natural world included – is not just for us, but for a panoply of non-human subjects as well. There is more to it than is revealed to us humans. And this is therefore one way that nature can disclose itself to us as a reality independent of human concerns. Yet there are others. For instance, nature can disclose itself as being *indifferent* to human concerns.

Indifferent in what sense? John Passmore writes that nature is not *'positively* indifferent' to human concerns, but *'incapable* of caring about us' (1995 138; cf. Taylor 1986: 122). But if one reads his statement as a universal claim about natural *beings* it is false. After all, some natural beings are sentient and so presumably capable of 'positive' indifference. One is occasionally justified in speaking of the positive indifference of animals, since it is reasonable to think that some of them are wrapped up in their own, peculiarly animal lives and so uninterested in the various things that concern us humans. In fact, it is likely that when we (us humans) do show up in animals' fields of significance, we do so only to the extent that we are, so to speak, labelled with instrumental values. The animal – the wild animal, certainly, and probably the domesticated animal for much of the time – is only concerned with us insofar as we affect or are likely to affect them. We register in their experience as prey, as predators, as providers of food or comfort and so forth.

So animals may be indifferent to us in the sense that although they care about some things, they typically don't care about us or about the things that concern us. However, many of those writers who draw attention to nature's indifference are not referring to this kind of indifference but to the way in which 'natural processes go on in their own way, in a manner indifferent to human interests' (Passmore 1995: 137), and this provides us with another way to understand nature's indifference.

But what exactly does it mean to say that nature is indifferent to human interests? In many cases, nature is thought to be indifferent, not to human interests and concerns *per se*, but to a certain subset of them. Thus part of the reason some disenchanted young people find themselves drawn to remote stretches of wilderness is because they regard such places as being indifferent to those concerns – with social standing or fashion or whatever – that characterise everyday – in their view, 'merely human' – existence. So, for instance, when he took off into the Alaskan wilderness, Chris McCandless seemed to be motivated primarily by a longing to escape a suburban life of 'security, conformity, and conservatism' (Krakauer 2007: 58).

Yet sometimes nature is thought to be indifferent, not just to some subset of human concerns, but to human concerns altogether. This kind of indifference is thought to be especially evident in one's dealings with harsh, wild places – baking deserts, windswept mountainsides and the like (see Abbey 1992: 267, for example, on desert landscapes). Yet one does not have to take to the wilderness to experience the indifference of nature. Walking along a beach, for instance, one might be struck by the sense that things in the beach-world – the hypnotic action of the waves, the gulls wheeling overhead, the slow reshaping of the cumulus on the horizon – are simply going on in their own ways, heedless of human concerns, not just concerns with social status and fashion, but human concerns *tout court*. To be sure, the beach-world might not be wholly unaffected by human concerns – for instance, the coastline might be to some extent managed – but insofar as it is natural, it will not reflect those concerns.

This sense of nature's indifference is beautifully conveyed in J. A. Baker's *The Peregrine*. The book is a record of one man's attempts to follow and observe peregrine falcons. And that is pretty much all there is to it. It is, as Robert Macfarlane explains,

> a book in which very little happens, over and over again. Dawn. The man watches, the bird hunts, the bird kills, the bird feeds. Dusk. And so on, through seven months.
>
> (Baker 2005: xii)

There is no plot to speak of. The reader isn't told where the action (such as it is) is taking place. Baker tells us nothing about himself, about who he is or what drives him to follow peregrines. And the author's indifference to literary conventions mirrors the indifference of his subject matter. For *The Peregrine* is a book from which humans and their works are eerily absent. A tractor in the distance, a ship out at sea: but other than this, the reader is presented with nothing but nature and natural things. It is a book about nature, directed by nature, running at its pace. Each passage rises with the sun and ends with its setting; we begin in October and end in April. And always at the centre the 'glazed inhuman' eye of the hawk (2005: 155), a creature whose 'manners', as Ted Hughes once put it, are 'tearing off heads' (2000: 17). The abiding impression is of a world that is utterly indifferent to human concerns.

And this indifference is reflected in the misanthropy of the book's narrator. As time passes, Baker becomes more and more like his quarry and increasingly removed from the sphere of human concerns. 'I sank into the skin and blood and bones of the hawk', he writes. 'Like the hawk, I heard and hated the sound of man, that faceless horror of the stony places' (2005: 144). For Baker, the natural world is indifferent to human concerns, yet for him this provides solace. In nature, in the pursuit of peregrines, he is able to slough off his human skin and re-establish some sort of connection with the natural world.

Φ

Nature's indifference to human concerns is especially evident in the loss of scale that natural phenomena can sometimes provoke. The most familiar instances of this involve encounters with unfathomably large 'things' – an expanse of sea or desert, the night sky, even, perhaps, a giant flock of birds (cf. Cocker 2008: 5). But a loss of scale can also be occasioned by a sense of the miniature. As Gaston Bachelard writes:

> Large issues from small, not through the logical law of a dialectics of contraries, but thanks to liberation from all obligations to dimensions. ... [T]he miniscule, a narrow gate, opens up an entire world. The details of a thing can be the sign of a new world which, like all worlds, contains the attributes of greatness.
>
> (1964: 154–5)

Thus, inspired by Bachelard, Macfarlane writes of being 'drawn into illusions in which my sense of scale was thrown, and it seemed that I might be able to enter a bird's nest, or the bole of a tree, or pass into the curled lustrous chamber of a whelk, following the whorl of its chamber round, keeping a hand pressed to its shiny surface, searching for its topmost spire' (2007: 251–2). To appreciate such things is to be privy to a world of which man is not the measure, a world that is in this sense indifferent to human concerns.

A sense of nature's indifference – and thus of its independent reality – can also be occasioned by a loss of *temporal* scale. Bachelard writes that '[w]e do not have to be long in the woods to experience the always rather anxious impression of "going deeper and deeper" into a limitless world' (1964: 185). Yet this depth is not merely spatial: '[W]ho knows the temporal dimensions of the forest?' he asks:

> History is not enough. We should have to know how the forest experiences its great age; why, in the reign of the imagination, there are no young forests
>
> (1964: 188)

The forest's past is not merely inferred; it is to some degree present in experience (see Rolston 2004: 182). In the phenomenologist's idiom, one could say that a sense of the past is integral to our perception of the forest, a horizon of it. We see the trees as rooted, and this perception is partly constituted by our sense of the roots beneath our feet, slowly pushing their way through the soil. And looking upwards, we know, too, how many years it has taken the tree to reach this far upwards towards the sun. Time here is congealed into the form of the trees – their leaning this way or that, the scars and furrows in their bark, the patterns of lichen-growth on their trunks. There are, as Richard Mabey puts it, 'layers of history' here, bound up in 'grain and forkings and slow cycles of light and shade' (2005: 11).

And if woods can awaken us to an unfathomable inhuman past, then so too – even more so – can geological features. To the trained eye, the slow uplift of mountains and the relentless inch-by-inch crawl of glaciers can open our senses to a still deeper time, and one that is, if anything, still further removed from the sphere of our all too human concerns. It is humbling to reflect that the land mass we now call the British Isles was millions of years ago a lush rainforest; millions of years later, the sun-bleached desert heart of the super-continent, Pangaea; and around 80 million years later, during the Jurassic, the silty bed of a sea stocked with icthyosaurs and ammonites. And looking forwards in time, rather than backwards, it's equally humbling to reflect that the slow shift of tectonic plates will continue long after we have vacated the planet and all our works have crumbled to dust.

5.3   The transhuman nature of perception

These are, I admit, just sketches. It would take a great deal more work to develop a comprehensive and organised account of the various ways in which the independent reality of nature can disclose itself to us in experience. Moreover, I do not mean to suggest that the examples I have cited above are all alike. In many important respects, sensing the alterity of a tree and perceiving an immense flock of crows (to recall just two of my examples) are quite different experiences of quite different phenomena. Nonetheless both are, in their different ways, experiences of the independent reality of nature, and as such they are both appropriate objects of phenomenological investigation.

These, then, are my initial responses to the charge of anthropocentrism: to concede that the phenomenological approach is focused on how the world discloses itself to beings like us, but to contend (1) that this does not amount to anthropocentrism (since the beings that are 'like us' in the relevant sense are not all human), and (2) that this focus may in fact be to its advantage and is, at any rate, something that ought not to perturb environmental thinkers. Precisely *because* they are focused on how the world discloses itself to beings like us, phenomenological inquiries have the potential to elucidate the various ways in which we encounter the independent reality of nature in the living of our lives. But there is more to be said here. For although the phenomenological account is, in one respect at least, focused on how the world discloses itself to beings like us, in another it need not be. Indeed – and this is my third response to the charge of anthropocentrism – a certain line of phenomenological inquiry, if followed through to its conclusion, results in a singularly *non*-anthropocentric account of the world. And this, I shall argue, is something that the environmental realists with whom we began would do well to note.

Φ

To see why and in what sense phenomenology might be non-anthropocentric, let us return to the question of what it means to be 'in-the-world'. To say that we are in-the-world is, we saw, to say that our way of being and that of the world are internally related, in the sense that the one cannot be understood without referring to the other. And to say this is, at least in part, to say that whatever shows up for us as a constituent of the world does so in relation to our concerns.

In *Being and Time*, Heidegger expresses this point by claiming that beings reveal themselves to us in the light of our moods and practical concerns. Similarly, in *Phenomenology of Perception* Merleau-Ponty conceives the world as the correlate of one's embodied subjectivity (Madison 1981: 27). But it would be wrong to infer, on this basis alone, that these thinkers' positions are at root anthropocentric. It is true that for both Heidegger and Merleau-Ponty whatever 'shows up' for us must do so in relation to our concerns, even if it reveals itself as being independent of those concerns. Yet in trying to understand being-in-the-world we are not exactly concerned with *what* shows up, but rather with the 'process' by which anything shows up at all. We are concerned with *presencing*, the process Heidegger often refers to as Be*ing*. And the relation of *this* to human faculties and concerns has not yet been made clear. Indeed, in the following I will set out some reasons for thinking that this process is to some extent *non*-human or *trans*human.

Φ

Let's begin with Heidegger. In *Being and Time*, Heidegger tries to explain what I have called 'presencing' in terms of the concept of a clearing (*Lichtung*). Dasein, he suggests, *is* such a clearing, by which he means that to be the sort of being each of us is, is not to be an item in the world (a material object, for example), but to be something like a space within which beings come to presence. This space is not mental in character: for Heidegger, the clearing is not a Cartesian realm of consciousness, within which beings disclose themselves as ideas. To be the sort of being each of us is, is to be in-the-world; hence the clearing within which beings present themselves is, so to speak, shared by humans and the world. To say that there is a human contribution here is to say that whatever shows up in the clearing does so in relation to our concerns, broadly construed. Yet the fact that Dasein is always being-in-*the-world* means that what shows up is not merely a product of our understanding. As Dorothea Frede puts it, '[t]he intelligibility [of the world] resides as much in the "things" encountered themselves as in the understanding residing in us'

(1993: 66). As we saw above, Heidegger therefore manages to espouse what might appear to be a certain kind of metaphysical anti-realism, but without endorsing the radically anthropocentric claim that the world is a product of human understanding. In rejecting the latter claim, Heidegger's account is to some degree non-anthropocentric.

Now all of this has been shown already. Here I would like to emphasise that Heidegger's account is non-anthropocentric in *two* respects: not just on account of his contention that what shows up in experience cannot be understood as a product of human understanding, but also because of his insistence that the very 'process' by which anything at all shows up is not entirely human in nature. Thus having noted that the fact that anything shows up in the clearing at all is as much a function of the beings encountered as of our understanding of them, Frede proceeds to remark that, in Heidegger's view, 'this "fittingness" [between beings and our understanding] is not due to any merit of ours' (1993: 66). Rather, it is something that 'simply happens to us, and in this sense "being" is quite out of our control ... an "opening", a "free gift"' (1993: 66).[9]

The upshot of these difficult ideas is that although the clearing is human-related, in that anything that shows up in it does so in relation to our concerns, it is not human-controlled (a point Heidegger came to emphasise in his later work). Or, to put things slightly differently: for Heidegger, the clearing is where Be-*ing* takes place, where Being is conceived as the 'showing up' of beings, their 'movement into presence' (see Guignon 1993: 13). But although what shows up does so in relation to our concerns, the 'process' whereby anything shows up is, to some extent at least, transhuman.

Φ

Merleau-Ponty, for his part, did not accept Heidegger's views on these matters. In his opinion, they were focused too much on Being, and not enough on human beings (see Madison 1981: 252). Yet despite this, there is, as early as *Phenomenology of Perception*, a discernibly non-anthropocentric aspect to Merleau-Ponty's thought, too.

In order to see this, recall Merleau-Ponty's claim, in *Phenomenology of Perception*, that perception is a kind of 'communion' (*PP*: 320). The thing, which seems so obviously to exist 'out there' in the world, beyond the limits of our consciousness, is, he says, partly constituted by our relations to it. '[I]ts articulations', he writes, 'are those of our very existence' (*PP*: 320).

Yet it is not obvious how this communion should be understood. It is not clear what elements are, so to speak, communing. In my earlier discussion of this issue I suggested, provisionally, that the communion is between the perceiver and the perceived. But that claim is problematic, for on Merleau-Ponty's account the relation between perceiver and perceived is so intimate as to militate against any attempt to conceive either element in isolation from the other. For this reason a communion between the two cannot be regarded as the combination of two elements that might have remained apart, even in thought. Perception is no more a communion of perceiver and perceived than a vase, say, is a combination of an outer surface and an inner one.

As we have seen, in *Phenomenology of Perception* the peculiar intimacy of perceiver and perceived is conveyed through a series of reflections on the role of the body in perception. The structure of our all-too-human experience does not constitute an epistemological barrier, debarring us access to how the world is 'in itself'. On the contrary, I can only perceive the world because I am not foreign to it, because I am 'made of the same stuff as it' (*N*: 218; cf. *PrP*: 16). I am able to perceive the world because I inhere in it. The claim here is not that one's body is merely a *means* by which one interacts with an external world. The relationship between perceiver and perceived is more intimate than that. And this is precisely what Merleau-Ponty means to convey in his various descriptions of how things disclose themselves in relation to one's body – as having a certain texture under the fingers, say, or as being within reach.

The general theme of the kinship between perceiver and perceived is continued in Merleau-Ponty's final, unfinished work, *The Visible and the Invisible*, and the famous discussion, in that work, of the experience of touching one hand with the other. The significant feature of such an experience, Merleau-Ponty claims, is that it is 'reversible'. It seems initially as though my right hand is touching my left, but then it is as if my right hand is itself being touched. Then, once again, it is as if my right is touching my left. And so on. An oscillation is set up, the hands, like two Sartrean individuals, playing alternately at being active and passive, yet neither at any single moment taking on both roles (cf. *PP*: 93).

Merleau-Ponty uses the example to make several points, only some of which need concern us here. First, he draws attention to the fact that the hand (either hand) is only able to touch because it is itself a touchable thing:

> [B]etween my movements and what I touch ... there must exist some ... kinship. ... This can happen only if my hand, while it is felt from within, is also accessible from without, itself tangible, for my other hand, for example, if it takes its place among the things it touches, is in a sense one of them ...
>
> (*VI*: 133)

These claims might seem compatible with some version of material-ism. But they are not. Merleau-Ponty's aim is not to postulate the existence of some material object (the body, say) of which one's hands are both parts, but to reflect on the conditions that must obtain in order for anything to reveal itself as an object in the first place. Moreover, while materialism is a theory, Merleau-Ponty does not present these thoughts as merely theoretical postulates. He is not *inferring* that some kinship exists between toucher and touched – he is suggesting that these conclusions are presented at some level in experience. Thus Merleau-Ponty maintains that the awareness of what it feels like to be touched enters into one's very experience of touching, such that one cannot fully comprehend the phenomenol-ogy of touching without understanding what it is to be touched. The two aspects are, he says, 'intertwined'.

The curious relation brought to light in the account of touch is taken to hold of other senses too. Consider vision, for instance.[10] On Merleau-Ponty's account, I would not be able to see if my eyes, oily, epispherical globes that they are, were not part of the visible world. Nor would anything show up in my visual field were I not the occupier of a certain location in the world, from which things radiate outwards in varying degrees of depth. I can only see, he maintains, because I am part of the see-able world: 'he who sees cannot possess the visible unless he is possessed by it, unless he *is of it*' (*VI*: 134–5). And as with touch, a phenomenological account of seeing is taken to refer to the possibility of being seen – Merleau-Ponty writes of the painter's sense that he is 'looked at' by what he paints (*VI*: 139; cf. 151) and of a state wherein 'the seer and the visible reciprocate one another and we no longer know which sees and which is seen' (*VI*: 139).[11]

The general picture here is of a peculiarly intimate relationship between perceiver and perceived, one more intimate than is suggested by talk of a 'communion' between the two. I do not stand over against the world as an active subject before a collection of merely passive objects (as was acknowledged in *Phenomenology of Perception*). Nor, however, is the world to be understood as the correlate of my embodied subjectivity (which, as Madison (1981: 27) notes, was how matters were conceived in that work). Rather, my perception of the world is, in truth, not mine at all – not entirely. I touch the thing, but at the same time I am myself susceptible, worldly, embodied being that I am, to what may be described as its touch (*VI*: 261). Better, then, to say that there is only a single moment of 'Tangibility', from which both man and thing are abstractions. Likewise with vision. It is not that I see the thing, nor even that the thing sees me, but that we, the thing and I, are brought into the fold of a wider dimension of 'Visibility'.

Merleau-Ponty tries to evoke this general relation between perceiver and perceived with the word 'flesh' (*la chair*). Just as the aspects of touching and being touched are 'intertwined', so he claims that the general bond between human being and world indicated by talk of 'perception' must be understood as a self-reflexive relation of a single 'element', a 'coiling over' or 'intertwining' of flesh.

There is clearly much more to be said here, but it is not my aim in this chapter to provide a comprehensive explanation, still less a defence, of Merleau-Ponty's notoriously difficult ontology of flesh (for a detailed account, see Madison 1981). For the purposes at hand, it will suffice to note that to conceive perception as a self-reflexive relation of a single 'element' is to picture it, not as a human act (cf. *VI*: 249), but as something like an event, and one, moreover, that has a transpersonal and indeed transhuman dimension. 'All flesh', Merleau-Ponty writes, 'radiates beyond itself' (*PrP*: 186) so that 'this flesh that one sees and touches is not all there is to flesh' (*VI*: 144). It envelopes not just one's own body but 'the whole of the sensible of which [that body] is part, and ... the world' (*VI*: 139; cf. *VI*: 142). Flesh exists also 'in other fields' (*VI*: 144) – it is at work, not only in what is usually called our perception of the world, but in the perceptions of certain non-human animals too. 'My' perception is in actuality an intertwining of flesh, yet the same may be said of canine or feline perception. Indeed, without this commonality, without this sharing of a common flesh, there could be no understanding between humans and dogs and cats.

These are abstruse thoughts, and it is unclear how, precisely, they should be understood. Yet even on this most cursory of sketches it should be clear that we are dealing here with an account of experience that is to some degree *non*-anthropocentric. So while Merleau-Ponty's account may appear to be human-centred in one respect, in another sense it definitely is not – and, moreover, the seeds of this non-anthropocentrism are evident as early as *Phenomenology of Perception* and the idea, mooted in that work, that perception is a kind of communion (cf. Madison 1981: 21).[12]

Φ

I do not mean to suggest that Heidegger's account of Being is equivalent to Merleau-Ponty's account of flesh. On the contrary, there are deep-rooted differences between them. Yet there are also some notable points of agreement.

First, both thinkers agree that whatever 'presences' necessarily does so in relation to our concerns, which is to say that something that bore absolutely no connection to our lived experience (like Merleau-Ponty's 'primitive nebula'), could not 'show up' for us as a being at all. So the world is in this respect traced out in terms of the concerns of beings like us. What is more, both thinkers agree that, were it not for the existence of such beings, nothing could 'presence'. (Heidegger maintains that Being 'needs' man (*QCT*: 40), Merleau-Ponty that the intertwining of flesh could not occur without our involvement (see further, Madison 1981: 252).) Nonetheless, both thinkers concur that this presencing is not *entirely* human, and that to a significant extent it is *trans*human (for Heidegger, an event of Being, for Merleau-Ponty, an intertwining of flesh).

In neither case do these claims about the 'transhuman' character of presencing contradict the apparently anti-realist claim that the world must be traced out in terms of the concerns of beings like us. Neither Heidegger nor Merleau-Ponty would countenance any suggestion that there is a way the world 'anyway is', entirely independent of our concerns. But there is no contradiction here, for to speak of Being or flesh is to refer neither to a world nor to a constituent of a world, but rather to a condition for the possibility of there being a world at all. The world might be indelibly 'human', but the fact that there is a world at all cannot be comprehended without referring to 'something' that transcends human concerns (see further, Cooper 2002).

Furthermore, although both Heidegger and Merleau-Ponty maintain that presencing is to some extent transhuman, neither thinker suggests that this conclusion can be arrived at *a priori*. Both insist that it is rather the fruit of an interrogation of *experience*. This is not to say (nonsensically) that presencing might present itself as an item of experience. It is rather to say that by attending to and reflecting on our experience as such, one can come to appreciate both the process of presencing and its transhuman nature.

Merleau-Ponty says very little about what it might be *like* to appreciate the transhuman character of presencing. He merely suggests that by interrogating experience one can come to appreciate the fact that perception is an intertwining of flesh. But Heidegger, for his part, has more to say on the issue. In essence, he suggests that attuning oneself to the transhuman nature of presencing involves the exercise of a particularly acute variety of attention. To be privy to the transhuman nature of presencing one must, on the one hand, have been released from certain second-order conceptions, one must have cultivated an 'openness to the mystery' (*DT*: 55). On the other hand, one must be released towards the thing, one must act as a 'shepherd of Being' (*BW*: 245), letting the thing show itself truly, according to its own possibilities and limitations. In this way, one comes to regard one's experience, not as an act of a subject, nor as a merely passive response to incoming sense data, but as a way of being open to that 'process' of Being on which the inextricable and inexplicable complex of human and world depends.

Φ

I have suggested that the phenomenological account can be reconciled with a certain kind of non-anthropocentrism. How does this relate to the position of the 'environmental realist'? Certainly, a Being- or flesh-centred realism (if it may be so described) would be quite unlike the positions advocated by Rolston and other environmental realists. This may be conceded; however, it must also be granted that the phenomenologists' respective accounts, to the extent that they are not human-centred, embody some of the non-anthropocentric sentiments that motivate realist accounts in the first place.

Imagine an old sycamore standing in a forest. A realist, were he contemplating the tree, might very well maintain that it exists in its own right, independent of human perspectives, attitudes, practical concerns and the like. He might, in other words, regard the more-than-human dimension of reality as existing out there, as it were, at the terminus of his experience. In his view, the sycamore, like Dr Johnson's rock, gives the lie to all anti-realist pretensions. Now both Heidegger and Merleau-Ponty endorse the realist's bare claim, that there is some aspect of reality that transcends the human. But they do not conceive this dimension as residing *in* the world. Instead, they maintain that it is at work in the *perception* of the world. So, to be sure, the phenomenologist would deny that the sycamore enjoys the kind of independent existence the realist attributes to it. Yet he would at the same time affirm that in placing my palm against the tree's bark or in looking up through its branches I am party to an event which is to a certain extent transhuman, a 'communion' of man and tree, or better, an event of Being or an intertwining of a flesh that 'traverses' me (*VI*: 140; cf. Abram 1996: 68).

Φ

Towards the beginning of this chapter, I quoted one environmental realist's call for 'a real appreciation of ... a world not limited by what we make of it' (Clark 1994: 125). We saw that despite her basic opposition to realism, the phenomenologist seeks to do justice to this thought. We occasionally see things as existing independently of us, as harbouring an 'earthy' or 'nonhuman' dimension. And this, I suggested, is particularly evident in our dealings with the natural world, the world of mountains, rivers and trees, rather than that of shopping malls and multi-storey car parks. Yet we also saw that the phenomenologist's acknowledgement of a more-than-human world might seem to be, at root, human-centred, and so at odds with the professed non-anthropocentrism of environmental realists. I responded to this charge in three ways: first, by suggesting that although the world necessarily reveals itself in the light of *our* concerns, the relevant community here includes at least some non-human beings; second, by arguing that the phenomenologist's focus on the experience of beings like us is not in fact a bad thing; and third, by noting that there is, in the work of Heidegger and Merleau-Ponty at least, a move towards a markedly non-anthropocentric account of human experience.

I admit that this discussion has been little more than a gesture towards some ideas that deserve a more thorough treatment. Yet, with these qualifications, our discussion of these ideas suggests a further conclusion regarding the environmental realism debate: that the environmental realist is in a sense right, but for the wrong reasons. For on the account sketched in the previous few sections, reality is indeed not limited by what we make of it, but this is not because we are able, as it were, to fix our gaze upon some part of it – the natural world, perhaps – that exists in its own right, independent of human attitudes, practical concerns and the like. The transhuman dimension of reality is at work in our gaze itself. Or rather, what we take to be our outlook on the world is in fact not ours at all, not entirely.

If true, these claims suggest that one cannot account for our inherence in a world not of our making simply by proposing that we exist in the midst of a realm that exists in its own right, independent of human concerns. Indeed, they suggest that environmental thinkers, if they are fully to appreciate the place of human beings in the larger scheme of things, would do well not to focus exclusively on the merits or otherwise of realism. For to remain within the confines of the realism versus anti-realism debate is to overlook the possibility that the transhuman dimension of reality is not so much there *in* the natural world, standing over against us onlooking subjects, as at work in the very disclosure of that world. And this, in turn, suggests that environmental thinkers, if they are to realise their non-anthropocentric ambitions, would do well to pay attention, not only to the transhuman world, but also to the transhuman nature of perception: not only, that is, to the mountains, rivers and trees, but to that transhuman event by which these things present themselves to us as the things they are. They would be well advised to consider an account of the more-than-human world opposed, not just to metaphysical realism, but to that broadly subject-centred conception of the world presupposed by realists and anti-realists alike.

Notes

1. Cf. Simon Critchley's account of the debate between Merleau-Ponty, A. J. Ayer and George Bataille over whether the sun existed before the appearance of *Homo sapiens* (2001: 36).
2. Later on, I will use a phenomenological approach to undercut the distinction between metaphysical realism and metaphysical anti-realism, as the two positions are typically conceived. But for the moment we can provisionally assume that the phenomenological position, as represented in the works of Heidegger and Merleau-Ponty, is anti-realist in the sense explained.
3. The claim is attributed to Hilary Putnam.
4. According to Hammond et al., this is not to say that the existence of the thing *reduces* to anticipations of further experiences. For Merleau-Ponty, they suggest, the thing is open to unending exploration, yet it also has a brute presence, a 'thereness' exceeding the sum of its possible appearances (1991: 198).
5. Cf. Michael Smith's discussion of the idea that wilderness 'offers us a chance to escape a world where all we see reflects "humanity" back at us' (1991: 152). My argument here mirrors that of the previous chapter. There, I argued that merely functional things tend to inhibit the development and exercise of attention. Here I am suggesting that since attention is required in order to discern the non-human element in things, it is more difficult to discern that element in one's dealings with merely functional things.
6. The 'nature' depicted by Cézanne is also 'inhuman' in the sense that it is not carved up into discrete and familiar objects externally related to one another, but is instead composed of things that are singularly ambiguous and fuzzy-edged (*CD*: 16; cf. Madison 1981: 79). On Merleau-Ponty's interpretation of the works of Cézanne and its philosophical significance, see Toadvine 1997.
7. Cf. Merleau-Ponty's claim, in 'Eye and Mind', that the fact that 'we are never in a position to take stock of everything objectively' may be regarded by some as a cause for regret. 'But this disappointment arises from [a] spurious fantasy. ... It is the regret of not being everything, and a rather groundless regret at that' (*PrP*: 189–90).
8. The links between Merleau-Ponty's sketches of an 'ontology from within' (*VI*: 237) and Schopenhauer's thought merit further investigation. For Schopenhauer, there is no possibility of encountering anything in the world that does not disclose itself according to the transcendental conditions brought into play in our experience. But for all this, one's experience itself, or more precisely, one's 'inner knowledge' of one's own will, is non-anthropocentric in the sense that the will of which we are aware is, at root, numerically identical with the Will that constitutes the true nature of all things. See further, Jacquette 2005: 82–92; also Toadvine 2003.

9. Cf. David E. Cooper's claim that, like Wittgenstein, Heidegger was inspired by a sense of 'wonder that we and the world are so fitted for one another that anything can, as it were, be a *something*, an identifiable thing present for us in thought and speech' (1997: 113).

10. Merleau-Ponty admits that his account would have to be changed in order to apply to vision: 'The eye cannot see the eye as the hand touches the other hand; it can be seen only in a mirror. The gap is larger between the seeing and the seen than between the touching and the touched' (*N*: 223). In the following sketch, I merely try to illustrate some of the similarities between Merleau-Ponty's treatment of touch and his treatment of vision. I will not discuss the difficulties one encounters in trying to relate the two.

11. Compare the following account from Annie Dillard: 'I saw the backyard cedar where the mourning doves roost charged and transfigured, each cell buzzing with flame. I stood on the grass with the lights in it, grass that was wholly fire, utterly focused and utterly dreamed. It was less like seeing than like being for the first time seen, knocked breathless by a powerful glance' (1998: 121).

12. See especially Merleau-Ponty's claims that 'I ought to say that *one* perceives in me, and not that I perceive', and that 'sensation necessarily appears to itself in a setting of generality ... [arising] from *sensibility* which has preceded it [*sic*]' such that it may be experienced as 'a modality of a general existence ... which runs though me without my being the cause of it' (*PP*: 215–6). Here in embryonic form is Merleau-Ponty's later conception of the flesh.

Conclusion

Let me begin this final section, not by reviewing what I have argued, but by considering how environmental ethics is usually done.

According to what I shall call the standard approach, normative ethics (environmental ethics included) proceeds as follows. One begins by identifying what it is that marks a being out as worthy of moral consideration. Then one determines which kinds of being possess it. If the property (or set of properties) in question is confined to humans and divine humanoids, one's ethic will be anthropocentric. If not, it will be, to some degree at least, non-anthropocentric.

Non-anthropocentric ethics of this kind come in a number of varieties. So, for instance, Peter Singer argues that a being is morally considerable if and only if it is capable of having interests, and he therefore regards any and all beings capable of having interests – that is, any and all sentient beings – as morally considerable (Singer 1993: Chapter 3). Paul Taylor, by contrast, suggests that a being is morally considerable if and only if it has a life that can go well or badly, and, by the lights of his particular brand of biocentrism, the moral circle therefore extends beyond the realm of sentient beings to include non-sentient living things such as trees and fungi (see Taylor 1986, 119*ff*.).

Adopting this general approach does not commit one to the truth of the implausible claim that there is only one morally relevant factor. So, for instance, one can consistently hold, as does Singer for instance, that our moral obligations regarding persons will differ from our obligations with respect to sentient non-persons. However, the fact remains that for proponents of this general approach there exists a single feature (or a single set of features) that marks a being out as worthy of moral consideration. If a being possesses this feature, it is thought to be worthy of moral consideration (though it may be worthy of special kinds of moral consideration for other reasons). If it lacks it, it is considered to be unworthy.

The basic assumption here is that one can identify a set of necessary and sufficient conditions for being an object of generalised moral consideration (O'Neill et al. 2008: 108). Yet in recent years this assumption has been challenged. For example, in their excellent introduction to environmental ethics, John O'Neill, Alan Holland and Andrew Light maintain that '[e]thical reflection needs to start from the plurality of relations and moral responses that are owed to beings – not from some generalised and underspecified concept of "moral consideration"' (2008: 110). More fully:

> The concept of moral concern or moral consideration in the abstract does not capture the variety of responses that are required of different objects under different descriptions. For example, Kant is right that particular capacities for rational reflection about ends demand a particular kind of respect from others. However, it does not follow that all other objects are mere things that fall out of the domain of proper regard. It is rather that the nature of that regard changes. Likewise, sentient beings demand from us a particular set of relations of benevolence that non-sentient beings cannot evoke – one cannot be cruel or kind to a carrot. However, it does not follow that there are no other sets of attitudes and responses that are owed to non-sentient living things in virtue of their nature.
>
> (2008: 109)

In the light of these thoughts, it might be tempting to reject talk of generalised moral concern or consideration and opt instead for some kind of moral pluralism. And, as I argued in Chapter 3, this would be the correct response.

Yet everything rests on what *kind* of pluralism one adopts. According to one kind of pluralism (not, however, the kind I endorse), the moral philosopher's first task is to work out what sorts of being exist in the world and to identify their intrinsic properties. Having done this, she needs to determine which of these properties are morally relevant, which, in other words, mark a being out as morally considerable. As a pluralist, she will identify more than one such property. So whereas a monist, like Singer, thinks that all those beings that are morally considerable can be accommodated within a single category, that of 'sentient beings', the pluralist will contend that a number of categories are required.

The interesting thing to note here is that this kind of pluralist and the monist agree on the method by which ethics must proceed. They agree that the epistemological task of determining what kinds of being there are with what kinds of properties can and should be carried out independently of ethical considerations. So one might begin by determining what kinds of being there are with what kinds of properties before moving on to ask which of those properties are morally relevant, or one might begin by asking, in a highly general way, what kinds of properties are morally relevant before moving on to ask how those properties are distributed in the world – but, either way, the epistemological part of the procedure is carried out without reference to ethical considerations: it is, one might say, purely epistemological, just as the ethical part is purely ethical.

If the case set out in the preceding pages is right, then this whole conception of the relation between epistemology and ethics is wrong. To suppose that the task of sorting different kinds of being into the appropriate ethical categories is *purely* epistemological is to say that it does not require the kind of morally significant attention to the world signalled by Murdoch's talk of 'moral vision' or Heidegger's references to releasement towards things. Indeed one might suppose further that to carry out such inquiries as if they were purely epistemological and void of any ethical significance is to be guilty of a lack of attention, a lack of moral vision. This, I take it, is Jim Cheney and Anthony Weston's point when they criticise 'epistemology-based ethics' for its failure to extend a 'basic courtesy' to the world, for its failure – to employ another idiom – to let beings be (1999: 116, 130). This basic lack of courtesy is, they suggest, evident in some of the ways people try to understand the moral status of animals. For, according to one popular conception, before one can arrive at an understanding of the moral status of animals, one must first identify what kinds of animal there are, with what properties. Once this purely epistemological task has been carried out, then one is thought to be in a position to determine which of those properties are morally significant (and how). But for Cheney and Weston, this is entirely wrong:

> On the usual view ... we must first know what animals are capable of, then decide on that basis whether and how we are to consider them ethically. On [our] view, we will have no idea of what other animals are actually capable – we will not readily understand them – until we have *already* approached them ethically – that is, until we have offered them the space and time, the occasion, and the acknowledgement necessary to enter into relationship. Ethics must come *first*.
>
> (1999: 118)

It is not clear whether Cheney and Weston mean to endorse Levinas' bold claim that any attempt to comprehend what *is* must cede its title as 'first philosophy' to ethics (see further, Levinas 1989). For my part, however, I would want to make the more moderate claim that neither epistemology nor ethics qualifies as 'first philosophy', but that the two approaches are instead coeval. There is no question of our determining what kinds of being there are in the world and only then turning to the ethical task of deciding how they ought to be treated. And there is no question of our engaging in 'pure ethics', abstracted from sensitive attention to how things are. Seeing the world clearly requires the development and exercise of moral virtue, while ethics, for its part, must be understood in terms of the exercise of what Murdoch calls selfless attention to nature. Seeing what *is* requires moral vision, while moral goodness involves what Murdoch refers to as 'obedience to reality' (2007: 41).

Φ

My central recommendation, then, is that attention is a moral virtue, which is to say that everyone – and not just phenomenologists – ought to be attentive. But I have also suggested that developing this virtue may be especially important for those of us who live in societies dominated by the forces of mass-production. And I have clarified and justified this claim by discussing what it means to attend to different kinds of object. So, for instance, in Chapter 2 I focused on our experience of animals. I hoped through my discussion to illustrate what it might mean to be attentive to the lives of animals. But I also hoped, by reflecting on the practices of those who live with animals, to shed light on what it might mean for anyone to be so attentive. In particular, I argued that investigating the mental life of an animal need not involve the amassing of evidence in support of the hypothesis that it is conscious. Instead it can involve one's learning to see the animal's behaviour in a certain way; perhaps, even, the forging of some sort of rapport with the creature. And this, I suggested, requires an attentive, open attitude on the part of the researcher, a willingness to regard the animal as at least potentially conscious – a 'basic courtesy', as Cheney and Weston would say.

In Chapter 3 I considered what it might mean to be attentive, not exactly to the world, but to the various ways the world matters to us. I suggested that things matter to us in a number of incommensurable ways, and that in this context attention therefore requires pluralism, an openness to these different sorts of 'mattering'. To comprehend the various ways things matter to us, I concluded, we need to work with a conceptual vocabulary rich enough to capture our bonds with non-human others, our needs for them (and theirs for us), and the various meanings they have for us. I argued that, in their preoccupation with environmental *values*, environmental ethicists tend to fail in this regard.

In Chapter 4 I suggested that our material circumstances typically encourage appetitive desires or cravings and so militate against the development and exercise of attention. Surrounded by what we take to be merely functional things, we find it hard to attend to the world. By contrast, natural things, precisely because they do not tend to reflect our cravings, can invite attention. And in this way they can foster the development of the virtue. Accordingly, I suggested that while, on the one hand, we ought to conserve physical nature through action, we also need, through attention, to conserve the meaning nature has for us as a realm independent of human practical concerns.

In Chapter 5 I considered the charge that phenomenological approaches are inherently anthropocentric and hence inherently at odds with environmental concerns. In response, I began by noting that phenomenologists are perfectly able to acknowledge the otherness or alterity of things, and I suggested, further, that they are able to show that this alterity is particularly evident in our dealings with specifically natural things. Whereas artefacts, especially mass-produced ones, tend to reflect our all too human concerns, natural things disclose themselves as harbouring what Merleau-Ponty calls a 'non-human element' (*PP*: 322). To illustrate this point, I considered two related examples of the alterity of nature: (1) the disclosure of natural things as objects for both human and non-human subjects, and (2) nature's disclosure as a realm indifferent to human concerns. In the third section I suggested that attention can be directed, not only towards *beings*, but also towards the 'process' of Being or presencing by which they disclose themselves to us as beings. To explain what this involves I examined Heidegger's writings and Merleau-Ponty's account of flesh.

Φ

My general aim in writing this book was to demonstrate the merits of a phenomenological approach to environmental philosophy. And in the preceding chapters I hope to have shown how such an approach can shed light on some of the key philosophical issues concerning the natural world and our relations to it. But the merits of phenomenology are not merely intellectual, which is to say that doing phenomenology is not just a matter of solving certain tricky intellectual puzzles. On the contrary, as Merleau-Ponty was fond of pointing out, the chief merit of phenomenology is that it can 'reawaken our experience of the world' (*PP*: 206). And in this respect – as Merleau-Ponty was also fond of pointing out – it is similar to art. Great works of phenomenology, like great works of art, embody the finely tuned attention of their creators. And for this reason they can open our eyes to the depth, richness and strangeness of the world we inhabit.

Something similar may be said of naturalists. J. A. Baker once said that '[t]he hardest thing of all to see is what is really there' (2005: 19). To be sure, he was writing as a naturalist, yet he might just as well have been writing as a phenomenologist or an artist. For, just as phenomenology embodies 'the same kind of attentiveness and wonder, the same demand for awareness' that one finds in the works of Balzac, Proust, Valéry and Cézanne (*PP*: xxi), so the naturalist must learn to let the natural world disclose itself under its own terms, according to its own possibilities and limitations. He must let it be, to recall Heidegger's apt phrase. Thus Paul Rezendes maintains that in order to track forest animals one must 'melt' into the forest; yet in order to do this, he adds, 'a quality of attention is necessary to allow what you normally think of as "you" to disappear' (1998: 71). In a similar vein, Mark Cocker writes of how for naturalists 'enquiring becomes a way of loving' (2008: 186) and how in the field 'the heart slows, the breath steadies, the senses attune to the moment, the worries of the world fall away' (2008: 138).

And just as the fine quality of the phenomenologist's attention is expressed in her works, or the artist's in stone, canvas or wood, so the naturalist's attention can find expression on paper. A winter's evening brought to life: 'our own falling star ... a lens of brightness on the southern horizon resting in its own bed of lemon and rose light'. Clouds looming overhead 'like icebergs in an ocean of cold winter blue' (Cocker 2008: 1). Baker's hawk, stooping, growing from 'a speck to a blur, to a bird, to a hawk, to a peregrine', a 'shark's head dropping from the sky'; beneath it, the sky 'shredded up ... torn by whirling birds', life 'conjured' from the still earth (2005: 178, 172, 42, 106). From Deakin, 'a lovely lush marsh orchid in intense purple and a huge stack of a flower like a wedding cake' (2008: 238). This is the thickness and richness of the natural world reborn in the thickness and richness of words.

What is brought out here is a world which from the start is shot through with lived significance. Of course, in order to be a world at all, any world must connect in some way to our lived experience. Anything we perceive, whether 'natural' or 'human', must disclose itself in relation to our concerns, and in this sense a certain, merely formal kind of parochialism is inescapable. Yet we can take solace in the thought that while the world must disclose itself in terms of our interests, attitudes, practical concerns and the like, we are not cut off from the rest of nature. For one thing, our outlook on the world is not ours alone, but is to some extent shared by our non-human cousins. For another, although we necessarily perceive a world lit up in terms of our concerns, we, fragile corporeal beings that we are, are only able to do this because we inhere in a world with which we are involved, because we are of the same 'flesh' as the things we perceive. Indeed, this perceiving is in a certain sense not *our* doing at all. It is the manifestation of nature working in us and through us – *natura naturans*, as Spinoza might have said, rather than *natura naturata*; nature as an active presence working behind the scenes, rather than the nature one finds represented in works of science or metaphysics. Nature, in this sense, is not something that we perceive; it is at work in our perception. It is not something we can capture in understanding, but something into which we are always already, and for the most part unknowingly, taken up.

Φ

These thoughts are difficult to sustain – particularly so in the buzz and bustle of everyday life. At work and at home, we are continually distracted by our material circumstances. The things that surround us call to us, reminding us not just of what we need to do, of what we fear, of what we regret, but also of our achievements, what we desire, what gives us comfort. And so we drift through life in a cloud of human concerns.

But in natural places it can be different. It is past three now on a December afternoon and I am strolling through a medium-sized beechwood a few miles from my home. The world seems calmer and quieter here, the surroundings less insistent. Now there is only *this* – not something grand and godlike, not Nature with a capital 'N', but this simple fragile moment. The muddy path, the bare branches overhead, the bite of the air, a clatter of pigeons in the canopy. These things, just as they are.

Bibliography

Aaltola, E. (2008) 'Personhood and Animals', *Environmental Ethics* 30 (2), pp. 175–93.

Abbey, E. (1992) *Desert Solitaire: A Season in the Wilderness* (London: Robin Clark Ltd).

Abram, D. (1996) *The Spell of the Sensuous: Perception and Language in a More-Than-Human World* (New York: Vintage Books).

Acampora, R. R. (2006) *Corporeal Compassion* (Pittsburgh: University of Pittsburgh Press).

Ansink, E.; Hein, L.; Per Hasund, K. (2008) 'To Value Functions or Services? An Analysis of Ecosystem Valuation Approaches', *Environmental Values* 17 (4): 489–503.

Bachelard, G. (1964) *The Poetics of Space*, trans. M. Jolas (New York: Orion Press).

Baker, G. and Morris, K. J. (1995) *Descartes' Dualism* (London: Routledge).

Baker, J. A. (2005) *The Peregrine*, introduction by Robert Macfarlane (New York: New York Review Book).

Banks, W. P. (1995) 'Evidence for Consciousness', *Consciousness and Cognition* 4, pp. 270–2.

Barbaras, R. (2001) 'Merleau-Ponty and Nature', *Research in Phenomenology* 31: 22–38.

Behnke, E. A. (1999) 'From Merleau-Ponty's Concept of Nature to an Interspecies Practice of Peace', in H. P. Steeves (ed.), *Animal Others: On Ethics, Ontology, and Animal Life* (Albany: SUNY Press), pp. 93–116.

Bekoff, M. (1996) 'Cognitive Ethology, Vigilance, Information Gathering, and Representation: Who Might Know What and Why?', *Behavioural Processes* 35, pp. 225–37.

Benso, S. (2000) *The Face of Things: A Different Side of Ethics* (Albany: SUNY Press).

Brady, E. (2003) *Aesthetics of the Natural Environment* (Edinburgh: Edinburgh University Press).

Brown, C. S. (2007) 'The Intentionality and Animal Heritage of Moral Experience', in C. Painter and C. Lotz (eds), *Phenomenology and the Non-Human Animal: At the Limits of Experience*, Contributions to Phenomenology 56 (Dordrecht: Springer), pp. 85–95.

Calarco, M. (2008) 'Heidegger's Zoontology', in P. Atterton and M. Calarco (eds), *Animal Philosophy: Ethics and Identity* (London: Continuum), pp. 18–30.

Caporael, L. R. (1986) 'Anthropomorphism and Mechanomorphism: Two Faces of the Human Machine', *Computers in Human Behavior* 2, pp. 215–34.

Caputo, J. D. (1993) *Demythologizing Heidegger* (Bloomington: Indiana University Press).

Carlson, A. (1998) 'Aesthetic Appreciation of the Natural Environment', in R. G. Botzler and S. J. Armstrong (eds), *Environmental Ethics: Divergence and Convergence* 2nd ed. (New York: McGraw-Hill), pp. 122–31.

Carman, T. (2008) *Merleau-Ponty* (Abingdon, Oxon: Routledge).

Cerbone, D. R. (2006) *Understanding Phenomenology* (Chesham: Acumen).

Cheney, J. and Weston, A. (1999) 'Environmental Ethics as Environmental Etiquette: Toward an Ethics-Based Epistemology', *Environmental Ethics* 21, pp. 115–34.

Churchill, S. (2008) 'Nature and Animality', in R. Diprose and J. Reynolds (eds), *Merleau-Ponty: Key Concepts* (Stocksfield: Acumen).

Clark, S. R. L. (1994) 'Global Religion', in R. Attfield and A. Belsey (eds), *Philosophy and the Natural Environment* (Cambridge: Cambridge University Press), pp. 113–28.

Clarke, M. (2002) 'Ontology, Ethics, and *Sentir*: Properly Situating Merleau-Ponty', *Environmental Values* 11, pp. 211–25.

Coates, P. (1998) *Nature: Western Attitudes since Ancient Times* (Cambridge: Polity Press).

Cocker, M. (2008) *Crow Country* (London: Vintage).

Cooper, D. E. (2002) *The Measure of Things: Humanism, Humility and Mystery* (Oxford: Oxford University Press).

Cooper, D. E. and James, S. P. (2005) *Buddhism, Virtue and Environment* (Aldershot: Ashgate).

—— (1997) 'Wittgenstein, Heidegger and Humility', *Philosophy* 72, pp. 105–23.

—— (1996) *Heidegger* (London: The Claridge Press).

—— (1992) 'The Idea of Environment', in D. E. Cooper and J. A. Palmer (eds), *The Environment in Question* (London: Routledge), pp. 165–80.

Corcoran, P. B. (1999) 'Formative Influences in the Lives of Environmental Educators in the United States', *Environmental Education Research* 5 (2), pp. 207–20.

Cottingham, J. (1978) '"A Brute to the Brutes?": Descartes' Treatment of Animals', *Philosophy* 53, pp. 551–61.

Critchley, S. (2001) *Continental Philosophy: A Very Short Introduction* (Oxford: Oxford University Press).

Curry, P. (2006) *Ecological Ethics: An Introduction* (Cambridge: Polity).

Dawkins, M. S. (1980) *Animal Suffering: The Science of Animal Welfare* (London: Chapman & Hall).

Deakin, R. (2008) *Wildwood: A Journey Through Trees* (London: Penguin).

Derrida, J. (2008) 'The Animal That Therefore I Am (More To Follow)', in M. Calarco and P. Atterton (eds), *Animal Philosophy: Essential Readings in Continental Thought* (London: Continuum), pp. 113–28.

—— (1989) *Of Spirit: Heidegger and the Question*, trans. G. Bennington and R. Bowlby (Chicago: University of Chicago Press).

Des Jardins, J. R. (2001) *Environmental Ethics: An Introduction to Environmental Philosophy* 3rd edn. (Belmont, California: Wadsworth).

Dillard, A. (1998) 'Seeing', in R. G. Botzler and S. J. Armstrong (eds), *Environmental Ethics: Divergence and Convergence* 2nd ed. (New York: McGraw-Hill), pp. 114–21.

Dreyfus, H. L. (1993) 'Heidegger on the Connection between Nihilism, Art, Technology, and Politics', in C. B. Guignon (ed.), *The Cambridge Companion to Heidegger* (Cambridge: Cambridge University Press), pp. 289–316.

Emad, P. (1977) 'Heidegger's Value-Criticism and its Bearing on the Phenomenology of Values', *Research in Phenomenology* 7, pp. 190–208.

Embree, L. (2008) 'A Beginning for the Phenomenological Theory of Primate Ethology', *Environmental Philosophy* V (1), pp. 61–74.

—— (1997) 'Problems of the Value of Nature in Phenomenological Perspective or What to Do about Snakes in the Grass', in J. G. Hart and L. Embree (eds), *Phenomenology of Values and Valuing* (Dordrecht: Kluwer), pp. 49–61.

Evernden, N. (1985) *The Natural Alien* (Toronto: University of Toronto Press).

Ferry, L. (2008) 'Neither Man nor Stone', in P. Atterton and M. Calarco (eds), *Animal Philosophy: Ethics and Identity* (London: Continuum), pp. 147–56.

Fields, W.; Savage-Rumbaugh, E. S.; Segerdahl, P. (2005) *Kanzi's Primal Language: The Cultural Initiation of Primates into Language* (Basingstoke: Palgrave Macmillan).

Foltz, B. (1995) *Inhabiting the Earth: Heidegger, Environmental Ethics, and the Metaphysics of Nature* (New Jersey: Humanities Press).

Fox, W. (1995) *Toward a Transpersonal Ecology* (Boston: Shambhala).

Frede, D. (1993) 'The Question of Being: Heidegger's Project', in C. Guignon (ed.), *The Cambridge Companion to Heidegger* (New York: Cambridge University Press), pp. 42–69.

Frodeman, R. (2006) 'The Policy Turn in Environmental Ethics', *Environmental Ethics* 28 (1), pp. 3–20.

Gaita, R. (2003) *The Philosopher's Dog* (London: Routledge).

Glendinning, S. (1998) *On Being With Others: Heidegger – Derrida – Wittgenstein* (London: Routledge).

Gowans, C. W. (2003) *Philosophy of the Buddha* (London: Routledge).

Grandin, T. and Johnson, C. (2006) *Animals in Translation: Using the Mysteries of Autism to Decode Animal Behavior* (Harcourt: Harvest Books).

Griffin, D. R. (1984) *Animal Thinking* (Cambridge: Harvard University Press).

Guignon, C. (1993) 'Introduction', in C. Guignon (ed.), *The Cambridge Companion to Heidegger* (New York: Cambridge University Press), pp. 1–41.

Guthrie, S. E. (1997) 'Anthropomorphism: A Definition and a Theory', in R. W. Mitchell, N. S. Thompson and H. Lyn Miles (eds), *Anthropomorphism, Anecdotes, and Animals* (Albany: SUNY Press), pp. 50–8.

Haar, M. (1987) *The Song of the Earth: Heidegger and the Grounds of the History of Being*, trans. Reginald Lilly (Bloomington: Indiana University Press).

Hammond, M.; Howarth, J.; Keat, R. (1991) *Understanding Phenomenology* (Oxford: Blackwell).

Hardy, T. (1986) *Far From the Madding Crowd* (Middlesex: Penguin Books).

Harrison, P. (1992) 'Descartes on Animals', *The Philosophical Quarterly* 42 (167), pp. 219–27.

Hatley, J. (2004) 'The Uncanny Goodness of Being Edible to Bears', in B. V. Foltz and R. Frodeman (eds), *Rethinking Nature: Essays in Environmental Philosophy* (Bloomington: Indiana University Press), pp. 13–31.

Heath, P. (1975) 'The Idea of a Phenomenological Ethics', in E. Piv evi (ed.), *Phenomenology and Philosophical Understanding* (Cambridge: Cambridge University Press), pp. 159–72.

Holland, A. (2007) 'The Value Space of Meaningful Relations', a paper presented at the 'Embodied Values' workshop held at the University of Edinburgh (5 July 2007).

Houellebecq, M. (2001) *Atomised*, trans. F. Wynne (London: Vintage).

Hughes, T. (2000) *Ted Hughes: Poems Selected by Simon Armitage* (London: Faber and Faber).

Hume, D. (2004) *An Enquiry Concerning the Principles of Morals* (New York: Prometheus Books).

Jacquette, D. (2005) *The Philosophy of Schopenhauer* (Chesham: Acumen).

James, S. P. (2009) 'Phenomenology and the Problem of Animal Minds', *Environmental Values* 18 (1), pp. 33–49.

—— (2007) 'Merleau-Ponty, Metaphysical Realism and the Natural World' *International Journal of Philosophical Studies* 16 (4), pp. 501–19.

—— (2006) 'Human Virtues and Natural Values', *Environmental Ethics* 28, pp. 339–54.

Jamieson, D. (1998) 'Science, Knowledge, and Animal Minds', *Proceedings of the Aristotelian Society* 98 (1), pp. 79–102.

Kennedy, J. S. (1992) *The New Anthropomorphism* (Cambridge: Cambridge University Press).

Kline, G. L. (1996) 'Meditations of a Russian Neo-Husserlian: Gustav Shpet's "The Skeptic and His Soul"', in B. R. Wachterhauser (ed.), *Phenomenology and Skepticism: Essays in Honour of James M. Edie* (Evanston: Northwestern University Press), pp. 144–63.

Kohák, E. (1984) *The Embers and the Stars* (Chicago: University of Chicago Press).

Krakauer, J. (2007) *Into the Wild* (London: Pan Macmillan).

Kupperus, G. (2007) 'Attunement, Deprivation, and Drive: Heidegger and Animality', in C. Painter and C. Lotz (eds), *Phenomenology and the Non-Human Animal: At the Limits of Experience*, Contributions to Phenomenology 56, Springer, 2007, pp. 13–28.

Lee, K. (1999) *The Natural and the Artefactual: The Implications of Deep Science and Deep Technology for Environmental Philosophy* (Lanham: Lexington Books).

Leopold, A. (1968) *A Sand County Almanac* (New York: Oxford University Press).

Levin, D. M. (1989) *The Listening Self: Personal Growth, Social Change and the Closure of Metaphysics* (London: Routledge).

Levinas, E. (1989) 'Ethics as First Philosophy', in S. Hand (ed.), *The Levinas Reader* (Oxford: Blackwell), pp. 75–87.

Mabey, R. (2005) *Nature Cure* (London: Chatto & Windus).

Macfarlane, R. (2007) *The Wild Places* (London: Granta Books).

MacNeice, L. (1979) *Collected Poems* (London: Faber & Faber).

Madison, G. B. (1981) *The Phenomenology of Merleau-Ponty: A Search for the Limits of Consciousness* (Athens, Ohio: Ohio University Press).

Matthews, E. (2002) *The Philosophy of Merleau-Ponty* (Chesham: Acumen).

McKibben, B. (1990) *The End of Nature* (London: Viking).

Merton, T. (1993) *Zen and the Birds of Appetite* (Boston: Shambhala).

Mitchell, R. W. (1997), 'Anthropomorphism and Anecdotes: A Guide for the Perplexed', in R. W. Mitchell, N. S. Thompson and H. Lyn Miles (eds), *Anthropomorphism, Anecdotes, and Animals* (Albany: SUNY Press), pp. 407–27.

Moran, D. (1999) *Introduction to Phenomenology* (London: Routledge).

Morgan, C. L. (1894) *Introduction to Comparative Psychology* (London: Scott).

Muir, J. (2007) *My First Summer in the Sierra*, introduction by Robert Macfarlane (Edinburgh: Canongate Books).

Murdoch, I. (2007) *The Sovereignty of Good* (Abingdon: Routledge).

—— (1993) *Metaphysics as a Guide to Morals* (London: Penguin).

Nagel, T. (1974) 'What Is It Like to Be a Bat?', *Philosophical Review* 83, pp. 435–50.

Neale, G. (1998) *The Green Travel Guide* (London: Earthscan Publications).

Nyanaponika Thera (1971) *The Power of Mindfulness: An Inquiry into the Scope of Bare Attention and the Principal Sources of its Strength* (Kandy, Ceylon: Buddhist Publication Society).

O'Neill, J.; Holland, A.; Light, A. (2008) *Environmental Values* (Abingdon: Routledge).

Painter, C. (2007) 'Appropriating the Philosophies of Edmund Husserl and Edith Stein: Animal Psyche, Empathy, and Moral Subjectivity', in C. Painter and C. Lotz (eds), *Phenomenology and the Non-Human Animal: At the Limits of Experience*, Contributions to Phenomenology 56, Springer, 2007, pp. 97–115.

Palmer, J. A.; Suggate, J.; Robottom, I.; Hart, P. (1999) 'Significant Life Experiences and Formative Influences on the Development of Adults' Environmental Awareness in the UK, Australia and Canada', *Environmental Education Research* 5 (2), pp. 181–200.

Palmer, J. A.; Suggate, J.; Bajd, B.; Hart, P.; Ho, R. K.; Ofwono-Orecho, J. K. W.; Peries, M.; Robottom, I.; Tsaliki, E.; Van Staden, C. (1999) 'An Overview of Significant Influences and Formative Experiences on the Developments of Adults' Environmental Awareness in Nine Countries', *Environmental Education Research*, 4 (4), pp. 445–64.

Passmore, J. (1995) 'Attitudes to Nature', in R. Elliot (ed.), *Environmental Ethics* (Oxford: Oxford University Press), pp. 129–41.

Polt, R. (1999) *Heidegger – An Introduction* (London: UCL Press).

Ratcliffe, M. (2007) *Rethinking Commonsense Psychology: A Critique of Folk Psychology, Theory of Mind and Simulation* (Basingstoke: Palgrave Macmillan).

Rezendes, P. (1998) *The Wild Within: Adventures in Nature and Animal Teachings* (New York: Tarcher/Putnam).

Rollin, B. E. (2007) 'Animal Mind: Science, Philosophy, and Ethics', *The Journal of Ethics* 11, pp. 253–74.

Rolston H. III (2004) 'The Aesthetic Experience of Forests', in A. Carlson and A. Berleant (eds), *The Aesthetics of Natural Environments* (Toronto: Broadview), pp. 182–96.

—— (2003) 'Value in Nature and the Nature of Value', in A. Light and H. Rolston III (eds), *Environmental Ethics: An Anthology* (Oxford: Blackwell), pp. 143–53.

—— (1997) 'Nature for Real: Is Nature a Social Construct'?, in T. D. J. Chappell (ed.), *The Philosophy of the Environment* (Edinburgh, Edinburgh University Press), pp. 38–64.

Romanes, G. J. (1882) *Animal Intelligence* (London: Kegan Paul).

Ruonakoski, C. (2007) 'Phenomenology and the Study of Animal Behavior', in C. Painter and C. Lotz (eds), *Phenomenology and the Non-Human Animal: At the Limits of Experience*, Contributions to Phenomenology 56, Springer, 2007, pp. 75–84.

Saito, Y. (2004) 'Appreciating Nature on Its Own Terms', in A. Carlson and A. Berleant (eds), *The Aesthetics of Natural Environments* (Toronto: Broadview), pp. 141–55.

Savage-Rumbaugh, S.; Shanker, S. G.; Taylor, T. J. (2001) *Apes, Language, and the Human Mind* (New York: Oxford University Press).

Schatzki, T. R. (1992) 'Early Heidegger on Being, The Clearing, and Realism', in H. Dreyfus and H. Hall (eds), *Heidegger: A Critical Reader* (Oxford: Blackwell), pp. 81–98.

Scheler, M. (1954) *The Nature of Sympathy*, trans. P. Heath with an introduction by W. Stark (London: Routledge & Kegan Paul).

Schopenhauer, A. (1969) *The World as Will and Representation* 1, trans. E. F. J. Payne (New York: Dover).

Shaner, D. E. (1989) 'The Japanese Experience of Nature', in J. B. Callicott and R. T. Ames (eds), *Nature in Asian Traditions of Thought* (Albany: SUNY Press), pp. 163–82.

Shapiro, K. (1997) 'A Phenomenological Approach to the Study of Nonhuman Animals', in R. W. Mitchell, N. S. Thompson and H. Lyn Miles (eds), *Anthropomorphism, Anecdotes, and Animals* (Albany: SUNY Press), pp. 277–95.

Shepard, P. (1995) 'Virtually Hunting Reality in the Forests of Simulacra', in M. E. Soulé and G. Lease (eds), *Reinventing Nature? Responses to Postmodern Deconstruction* (Washington DC: Island Press), pp. 17–29.

Singer, P. (1993) *Practical Ethics*, 2nd ed. (Cambridge: Cambridge University Press).

Smith, M. F. (1991) 'Letting in the Jungle', *Journal of Applied Philosophy* 8 (2), pp. 145–54.

Smith, R. L. and Smith, T. M. (2001) *Ecology & Field Biology* (6th ed.) (San Francisco, CA: Benjamin Cummings).

Smuts, B. (1999) 'Reflections', in J. M. Coetzee's *The Lives of Animals* A. Gutmann (ed.), (Princeton, New Jersey: Princeton University Press), pp. 107–20.

Sober, E. (2002) 'Philosophical Problems for Environmentalism', in D. Schmidtz and E. Willott (eds), *Environmental Ethics: What Really Matters; What Really Works* (New York: Cambridge University Press), pp. 145–57.

Spada, E. C. (1997) 'Amorphism, Mechanomorphism, and Anthropomorphism', in R. W. Mitchell, N. S. Thompson and H. Lyn Miles (eds), *Anthropomorphism, Anecdotes, and Animals* (Albany: SUNY Press), pp. 37–49.

Spengler, O. (1960) *Man and Technics: A Contribution to the Philosophy of Life* (Connecticut: Greenwood Press).

Spinoza, B. (1996) *Ethics*, trans. E. Curley (London: Penguin).

Swanton, C. (2003) *Virtue Ethics: A Pluralistic View* (Oxford: Oxford University Press).

Taylor, P. (1986) *Respect for Nature: A Theory of Environmental Ethics* (Princeton: Princeton University Press).

Thomson, I. (2005) *Heidegger on Ontotheology: Technology and the Politics of Education* (New York: Cambridge University Press).

—— (2004) 'Ontology and Ethics at the Intersection of Phenomenology and Environmental Philosophy', *Inquiry* 47, pp. 380–412.

Thoreau, H. D. (1999) *Walden*, S. Fender (ed.) (Oxford: Oxford University Press).

Toadvine, T. (2003) 'The Primacy of Desire and Its Ecological Consequences', in C. S. Brown and T. Toadvine (eds), *Eco-Phenomenology: Back to the Earth Itself* (Albany: State University of New York Press), pp. 139–53.

—— (1997) 'The Art of Doubting: Merleau-Ponty and Cézanne', *Philosophy Today* 41, pp. 545–53.

Urry, J. (1995) *Consuming Places* (London: Routledge).

Vogel, S. (2003) 'The Nature of Artifacts', *Environmental Ethics* 25 (2), pp. 149–68.

Walker, A. (1994) 'The Place Where I was Born', in M. Walker (ed.), *Reading the Environment* (New York: W.W. Norton and Co.), pp. 94–8.

Warren, K. (2001) 'The Power and the Promise of Ecological Feminism' in L. Pojman (ed.), *Environmental Ethics: Readings in Theory and Application* (Belmont, California: Wadsworth), pp. 189–99.

Williams, B. (1992) 'Must a Concern for the Environment be Centred on Human Beings?', in C. C. W. Taylor (ed.), *Ethics and the Environment* (Didcot: Bocardo Press), pp. 60–8.

—— (1985) *Ethics and the Limits of Philosophy* (London: Fontana).

Witoszek, N. and Brennan, A. (eds), (1999) *Philosophical Dialogues: Arne Naess and the Progress of Ecophilosophy* (Lanham, Maryland: Rowan & Littlefield).

Wittgenstein, L. (1969) *On Certainty*, trans. D. Paul and G. E. M. Anscombe (Oxford: Basil Blackwell).

Woolf, L. (1981) *The Village in the Jungle* (New Delhi: Oxford University Press).

Wordsworth, W. (1994) *Selected Poems*, J. O. Hayden (ed.), (London: Penguin Classics).

Young, J. (2008) *Heidegger's Later Philosophy* (Cambridge: Cambridge University Press).

Zahavi, D. (2001) 'Beyond Empathy: Phenomenological Approaches to Intersubjectivity', in E. Thompson (ed.), *Between Ourselves: Second-Person Issues in the Study of Consciousness* (Thorverton: Imprint Academic).

Index